The Nikola Tesla Collection:

My Inventions, The True Wireless, Talking with the Planets, the Problem of Increasing Human Energy, On Light and Other High Frequency Phenomena

My Inventions

Nikola Tesla's Autobiography

At the age of 63 Tesla tells the story of his creative life.

First published in 1919 in the Electrical Experimenter magazine

Table of Contents

I.   My Early Life

The progressive development of man is vitally dependent on invention. It is the most important product of his creative brain. Its ultimate purpose is the complete mastery of mind over the material world, the harnessing of the forces of nature to human needs. This is the difficult task of the inventor who is often misunderstood and unrewarded. But he finds ample compensation in the pleasing exercises of his powers and in the knowledge of being one of that exceptionally privileged class without whom the race would have long ago perished in the bitter struggle against pitiless elements.

Speaking for myself, I have already had more than my full measure of this exquisite enjoyment, so much that for many years my life was little short of continuous rapture. I am credited with being one of the hardest workers and perhaps I am, if thought is the equivalent of labor, for I have devoted to it almost all of my waking hours. But if work is interpreted to be a definite performance in a specified time according to a rigid rule, then I may be the worst of idlers. Every effort under compulsion demands a sacrifice of life-energy. I never paid such a price. On the contrary, I have thrived on my thoughts.

In attempting to give a connected and faithful account of my activities in this series of articles which will be presented with the assistance of the Editors of the ELECTRICAL EXPERIMENTER and are chiefly addrest to our young men readers, I must dwell, however reluctantly, on the impressions of my youth and the circumstances and events which have been instrumental in determining my career.

Our first endeavors are purely instinctive, promptings of an imagination vivid and undisciplined. As we grow older reason asserts itself and we become more and more systematic and designing. But those early impulses, tho not immediately productive, are of the greatest moment and may shape our very destinies. Indeed, I feel now that had I understood and cultivated instead of suppressing them, I would have added substantial value to my bequest to the world. But not until I had attained manhood did I realize that I was an inventor.

This was due to a number of causes. In the first place I had a brother who was gifted to an extraordinary degree — one of those rare phenomena of mentality which biological investigation has failed to explain. His premature death left my parents disconsolate. We owned a horse which had been presented to us by a dear friend. It was a magnificent animal of Arabian breed, possest of almost human intelligence, and was cared for and petted by the whole family, having on one occasion saved my father's life under remarkable circumstances. My father had been called one winter night to perform an urgent duty and while crossing the mountains, infested by wolves, the horse became frightened and ran away, throwing him violently to the ground. It arrived home bleeding and exhausted, but after the alarm was sounded immediately dashed off again, returning to the spot, and before the searching party were far on the way they were met by my father, who had recovered consciousness and remounted, not realizing that he had been lying in the snow for several hours. This horse was responsible for my brother's injuries from which he died. I witnest the tragic scene and altho fifty-six years have elapsed since, my visual impression of it has lost none of its force. The recollection of his attainments made every effort of mine seem dull in comparison.

Anything I did that was creditable merely caused my parents to feel their loss more keenly. So I grew up with little confidence in myself. But I was far from being considered a stupid boy, if I am to judge from an incident of which I have still a strong remembrance. One day the Aldermen were passing thru a street where I was at play with other boys. The oldest of these venerable gentlemen — a wealthy citizen — paused to give a silver piece to each of us. Coming to me he suddenly stopt and commanded, "Look in my eyes." I met his gaze, my hand outstretched to receive the much valued coin, when, to my dismay, he said, "No, not much, you can get nothing from me, you are too smart." They used to tell a funny story about me. I had two old aunts with wrinkled faces, one of them having two teeth protruding like the tusks of an elephant which she buried in my cheek every time she kist me. Nothing would scare me more than the prospect of being hugged by these as affectionate as unattractive relatives. It happened that while being carried in my mother's arms they asked me who was the prettier of the two. After examining their faces intently, I answered thoughtfully, pointing to one of them, "This here is not as ugly as the other."

Then again, I was intended from my very birth for the clerical profession and this thought constantly opprest me. I longed to be an engineer but my father was inflexible. He was the son of an officer who served in the army of the Great Napoleon and, in common with his brother, professor of mathematics in a prominent institution, had received a military education but, singularly enough, later embraced the clergy in which vocation he achieved eminence. He was a very erudite man, a veritable natural philosopher, poet and writer and his sermons were said to be as eloquent as those of Abraham a Sancta-Clara. He had a prodigious memory and frequently recited at length from works in several languages. He often remarked playfully that if some

6

of the classics were lost he could restore them. His style of writing was much admired. He penned sentences short and terse and was full of wit and satire. The humorous remarks he made were always peculiar and characteristic. Just to illustrate, I may mention one or two instances. Among the help there was a cross-eyed man called Mane, employed to do work around the farm. He was chopping wood one day. As he swung the axe my father, who stood nearby and felt very uncomfortable, cautioned him, "For God's sake, Mane, do not strike at what you are looking but at what you intend to hit." On another occasion he was taking out for a drive a friend who carelessly permitted his costly fur coat to rub on the carriage wheel. My father reminded him of it saying, "Pull in your coat, you are ruining my tire." He had the odd habit of talking to himself and would often carry on an animated conversation and indulge in heated argument, changing the tone of his voice. A casual listener might have sworn that several people were in the room.

Altho I must trace to my mother's influence whatever inventiveness I possess, the training he gave me must have been helpful. It comprised all sorts of exercises — as, guessing one another's thoughts, discovering the defects of some form or expression, repeating long sentences or performing mental calculations. These daily lessons were intended to strengthen memory and reason and especially to develop the critical sense, and were undoubtedly very beneficial.

My mother descended from one of the oldest families in the country and a line of inventors. Both her father and grandfather originated numerous implements for household, agricultural and other uses. She was a truly great woman, of rare skill, courage and fortitude, who had braved the storms of life and past thru many a trying experience. When she was sixteen a virulent pestilence swept the country. Her father was called away to administer the last sacraments to the dying and during his absence she went alone to the assistance of a neighboring family who were stricken by the dread disease. All of the members, five in number, succumbed in rapid succession. She bathed, clothed and laid out the bodies, decorating them with flowers according to the custom of the country and when her father returned he found everything ready for a Christian burial. My mother was an inventor of the first order and would, I believe, have achieved great things had she not been so remote from modern life and its multifold opportunities. She invented and constructed all kinds of tools and devices and wove the finest designs from thread which was spun by her. She even planted the seeds, raised the plants and separated the fibers herself. She worked indefatigably, from break of day till late at night, and most of the wearing apparel and furnishings of the home was the product of her hands. When she was past sixty, her fingers were still nimble enough to tie three knots in an eyelash.

There was another and still more important reason for my late awakening. In my boyhood I suffered from a peculiar affliction due to the appearance of images, often accompanied by strong flashes of light, which marred the sight of real objects and interfered with my thought and action. They were pictures of things and scenes which I had really seen, never of those I imagined. When a word was spoken to me the image of the object it designated would present itself vividly to my vision and sometimes I was quite unable to distinguish whether what I saw was tangible or not. This caused me great discomfort and anxiety. None of the students of psychology or physiology whom I have consulted could ever explain satisfactorily these phenomena. They seem to have been unique altho I was probably predisposed as I know that my brother experienced a similar trouble. The theory I have formulated is that the images were the result of a reflex action from the brain on the retina under great excitation. They certainly were not hallucinations such as are produced in diseased and anguished minds, for in other respects I was normal and composed. To give an idea of my distress, suppose that I had witnest a funeral or some such nerve-racking spectacle. Then, inevitably, in the stillness of night, a vivid picture of the scene would thrust itself before my eyes and persist despite all my efforts to banish it. Sometimes it would even remain fixt in space tho I pushed my hand thru it. If my explanation is correct, it should be able to project on a screen the image of any object one conceives and make it visible. Such an advance would revolutionize all human relations. I am convinced that this wonder can and will be accomplished in time to come; I may add that I have devoted much thought to the solution of the problem.

To free myself of these tormenting appearances, I tried to concentrate my mind on something else I had seen, and in this way I would of ten obtain temporary relief; but in order to get it I had to conjure continuously new images. It was not long before I found that I had exhausted all of those at my command; my "reel" had run out, as it were, because I had seen little of the world — only objects in my home and the immediate surroundings. As I performed these mental operations for the second or third time, in order to chase the appearances from my vision, the remedy gradually lost all its force. Then I instinctively commenced to make excursions beyond the limits of the small world of which I had knowledge, and I saw new scenes. These were at first very blurred and indistinct, and would flit away when I tried to concentrate my attention upon them, but by and by I succeeded in fixing them; they gained in strength and distinctness and finally assumed the concreteness of real things. I soon discovered that my best comfort was attained if I simply went on in my vision farther and farther, getting new impressions all the time, and so I began to travel — of course, in my mind. Every night (and sometimes during the day), when alone, I would start on my journeys — see new places, cities and countries — live there, meet people and make friendships and acquaintances and, however unbelievable, it is a fact that they were just as dear to me as those in actual life and not a bit less intense in their manifestations.

This I did constantly until I was about seventeen when my thoughts turned seriously to invention. Then I observed to my delight that I could visualize with the greatest facility. I needed no models, drawings or experiments. I could picture them all as real in my mind. Thus I have been led unconsciously to evolve what I consider a new method of materializing inventive concepts and ideas, which is radically opposite to the purely experimental and is in my opinion ever so much more expeditious and efficient. The moment one constructs a device to carry into practise a crude idea he finds himself unavoidably engrost with the details and defects of the apparatus. As he goes on improving and reconstructing, his force of concentration diminishes and he loses sight of the great underlying principle. Results may be obtained but always at the sacrifice of quality.

My method is different. I do not rush into actual work. When I get an idea I start at once building it up in my imagination. I change the construction, make improvements and operate the device in my mind. It is absolutely immaterial to me whether I run my turbine in thought or test it in my shop. I even note if it is out of balance. There is no difference whatever, the results are the same. In this way I am able to rapidly develop and perfect a conception without touching anything. When I have gone so far as to embody in the invention every possible improvement I can think of and see no fault anywhere, I put into concrete form this final product of my brain. Invariably my device works as I conceived that it should, and the experiment comes out exactly as I planned it. In twenty years there has not been a single exception. Why should it be otherwise? Engineering, electrical and mechanical, is positive in results. There is scarcely a subject that cannot be mathematically treated and the effects calculated or the results determined beforehand from the available theoretical and practical data. The carrying out into practise of a crude idea as is being generally done is, I hold, nothing but a waste of energy, money and time.

My early affliction had, however, another compensation. The incessant mental exertion developed my powers of observation and enabled me to discover a truth of great importance. I had noted that the appearance of images was always preceded by actual vision of scenes under peculiar and generally very exceptional conditions and I was impelled on each occasion to locate the original impulse. After a while this effort grew to be almost automatic and I gained great facility in connecting cause and effect. Soon I became aware, to my surprise, that every thought I conceived was suggested by an external impression. Not only this but all my actions were prompted in a similar way. In the course of time it became perfectly evident to me that I was merely an automaton endowed with power of movement, responding to the stimuli of the sense organs and thinking and acting accordingly. The practical result of this was the art of telautomatics which has been so far carried out only in an imperfect manner. Its latent possibilities will, however, be eventually shown. I have been since years planning self-controlled automata and believe that mechanisms can be produced which will act as if possest of reason, to a limited degree, and will create a revolution in many commercial and industrial departments.

I was about twelve years old when I first succeeded in banishing an image from my vision by wilful effort, but I never had any control over the flashes of light to which I have referred. They were, perhaps, my strangest experience and inexplicable. They usually occurred when I found myself in a dangerous or distressing situation, or when I was greatly exhilarated. In some instances I have seen all the air around me filled with tongues of living flame. Their intensity, instead of diminishing, increased with time and seemingly attained a maximum when I was about twenty-five years old. While in Paris, in 1883, a prominent French manufacturer sent me an invitation to a shooting expedition which I accepted. I had been long confined to the factory and the fresh air had a wonderfully invigorating effect on me. On my return to the city that night I felt a positive sensation that my brain had caught fire. I saw a light as tho a small sun was located in it and I past the whole night applying cold compressions to my tortured head. Finally the flashes diminished in frequency and force but it took more than three weeks before they wholly subsided. When a second invitation was extended to me my answer was an emphatic NO!

These luminous phenomena still manifest themselves from time to time, as when a new idea opening up possibilities strikes me, but they are no longer exciting, being of relatively small intensity. When I close my eyes I invariably observe first, a background of very dark and uniform blue, not unlike the sky on a clear but starless night. In a few seconds this field becomes animated with innumerable scintillating flakes of green, arranged in several layers and advancing towards me. Then there appears, to the right, a beautiful pattern of two systems of parallel and closely spaced lines, at right angles to one another, in all sorts of colors with yellow -green and gold predominating. Immediately thereafter the lines grow brighter and the whole is thickly sprinkled with dots of twinkling light. This picture moves slowly across the field of vision and in about ten seconds vanishes to the left, leaving behind a ground of rather unpleasant and inert grey which quickly gives way to a billowy sea of clouds, seemingly trying to mould themselves in living shapes. It is curious that I cannot project a form into this grey until the second phase is reached. Every time, before falling asleep, images of persons or objects flit before my view. When I see them I know that I am about to lose consciousness. If they are absent and refuse to come it means a sleepless night.

To what an extent imagination played a part in my early life I may illustrate by another odd experience. Like most children I was fond of jumping and developed an intense desire to support myself in the air. Occasionally a strong wind richly charged with oxygen blew from the mountains rendering my body as light as cork and then I would leap and float in space for a long time. It was a delightful sensation and my disappointment was keen when later I undeceived myself.

During that period I contracted many strange likes, dislikes and habits, some of which I can trace to external impressions while others are unaccountable. I had a violent aversion against the earrings of women but other ornaments, as bracelets, pleased me more or less according to design. The sight of a pearl would almost give me a fit but I was fascinated with the glitter of crystals or objects with sharp edges and plane surfaces. I would not touch the hair of other people except, perhaps, at the point of a revolver. I would get a fever by looking at a peach and if a piece of camphor was anywhere in the house it caused me the keenest discomfort. Even now I am not insensible to some of these upsetting impulses. When I drop little squares of paper in a dish filled with liquid, I always sense a peculiar and awful taste in my mouth. I counted the steps in my walks and calculated the cubical contents of soup plates, coffee cups and pieces of food —

otherwise my meal was unenjoyable. All repeated acts or operations I performed had to be divisible by three and if I mist I felt impelled to do it all over again, even if it took hours.

Up to the age of eight years, my character was weak and vacillating. I had neither courage or strength to form a firm resolve. My feelings came in waves and surges and vibrated unceasingly between extremes. My wishes were of consuming force and like the heads of the hydra, they multiplied. I was opprest by thoughts of pain in life and death and religious fear. I was swayed by superstitious belief and lived in constant dread of the spirit of evil, of ghosts and ogres and other unholy monsters of the dark. Then, all at once, there came a tremendous change which altered the course of my whole existence. Of all things I liked books the best. My father had a large library and whenever I could manage I tried to satisfy my passion for reading. He did not permit it and would fly into a rage when he caught me in the act. He hid the candles when he found that I was reading in secret. He did not want me to spoil my eyes. But I obtained tallow, made the wicking and cast the sticks into tin forms, and every night I would bush the keyhole and the cracks and read, often till dawn, when all others slept and my mother started on her arduous daily task. On one occasion I came across a novel entitled "Abafi" (the Son of Aba), a Serbian translation of a well known Hungarian writer, Josika. This work somehow awakened my dormant powers of will and I began to practise self-control. At first my resolutions faded like snow in April, but in a little while I conquered my weakness and felt a pleasure I never knew before — that of doing as I willed. In the course of time this vigorous mental exercise became second nature. At the outset my wishes had to be subdued but gradually desire and will grew to be identical. After years of such discipline I gained so complete a mastery over myself that I toyed with passions which have meant destruction to some of the strongest men. At

14

a certain age I contracted a mania for gambling which greatly worried my parents. To sit down to a game of cards was for me the quintessence of pleasure. My father led an exemplary life and could not excuse the senseless waste of time and money in which I indulged. I had a strong resolve but my philosophy was bad. I would say to him, "I can stop whenever I please but is it worth while to give up that which I would purchase with the joys of Paradise?" On frequent occasions he gave vent to his anger and contempt but my mother was different. She understood the character of men and knew that one's salvation could only be brought about thru his own efforts. One afternoon, I remember, when I had lost all my money and was craving for a game, she came to me with a roll of bills and said, "Go and enjoy yourself. The sooner you lose all we possess the better it will be. I know that you will get over it." She was right. I conquered my passion then and there and only regretted that it had not been a hundred times as strong. I not only vanquished but tore it from my heart so as not to leave even a trace of desire. Ever since that time I have been as indifferent to any form of gambling as to picking teeth.

During another period I smoked excessively, threatening to ruin my health. Then my will asserted itself and I not only stopt but destroyed all inclination. Long ago I suffered from heart trouble until I discovered that it was due to the innocent cup of coffee I consumed every morning. I discontinued at once, tho I confess it was not an easy task. In this way I checked and bridled other habits and passions and have not only preserved my life but derived an immense amount of satisfaction from what most men would consider privation and sacrifice.

After finishing the studies at the Polytechnic Institute and University I had a complete nervous breakdown and while the malady lasted I observed many phenomena strange and unbelievable.

II. My First Efforts At Invention

I shall dwell briefly on these extraordinary experiences, on account of their possible interest to students of psychology and physiology and also because this period of agony was of the greatest consequence on my mental development and subsequent labors. But it is indispensable to first relate the circumstances and conditions which preceded them and in which might be found their partial explanation.

From childhood I was compelled to concentrate attention upon myself. This caused me much suffering but, to my present view, it was a blessing in disguise for it has taught me to appreciate the inestimable value of introspection in the preservation of life, as well as a means of achievement. The pressure of occupation and the incessant stream of impressions pouring into our consciousness thru all the gateways of knowledge make modern existence hazardous in many ways. Most persons are so absorbed in the contemplation of the outside world that they are wholly oblivious to what is passing on within themselves.

The premature death of millions is primarily traceable to this cause. Even among those who exercise care it is a common mistake to avoid imaginary, and ignore the real dangers. And what is true of an individual also applies, more or less, to a people as a whole. Witness, in illustration, the prohibition movement. A drastic, if not unconstitutional, measure is now being put thru in this country to prevent the consumption of alcohol and yet it is a positive fact that coffee, tea, tobacco, chewing gum and other stimulants, which are freely indulged in even at the tender age, are vastly more injurious to the national body, judging from the number of those who succumb. So, for instance, during my student years I gathered from the published necrologues in Vienna, the home of coffee drinkers, that deaths from heart trouble sometimes reached sixty-seven per cent of the total. Similar observations might probably be made in cities where the consumption of tea is excessive. These delicious beverages superexcite and gradually exhaust the fine fibers of the brain. They also interfere seriously with arterial circulation and should be enjoyed all the more sparingly as their deleterious effects are slow and imperceptible. Tobacco, on the other hand, is conducive to easy and pleasant thinking and detracts from the intensity and concentration necessary to all original and vigorous effort of the intellect. Chewing gum is helpful for a short while but soon drains the glandular system and inflicts irreparable damage, not to speak of the revulsion it creates. Alcohol in small quantities is an excellent tonic, but is toxic in its action when absorbed in larger amounts, quite immaterial as to whether it is taken in as whiskey or produced in the stomach from sugar. But it should not be overlooked that all these are great eliminators assisting Nature, as they do, in upholding her stern but just law of the survival of the fittest. Eager reformers should also be mindful of the eternal perversity of mankind which makes the indifferent "laissez-faire" by far preferable to enforced restraint.

The truth about this is that we need stimulants to do our best work under present living conditions, and that we must exercise moderation and control our appetites and inclinations in every direction. That is what I have been doing for many years, in this way maintaining myself young in body and mind. Abstinence was not always to my liking but I find ample reward in the agreeable experiences I am now making. Just in the hope of converting some to my precepts and convictions I will recall one or two.

A short time ago I was returning to my hotel. It was a bitter cold night, the ground slippery, and no taxi to be had. Half a block behind me followed another man, evidently as anxious as myself to get under cover. Suddenly my legs went up in the air. In the same instant there was a flash in my brain, the nerves responded, the muscles contracted, I swung thru 180 degrees and landed on my hands. I resumed my walk as tho nothing had happened when the stranger caught up with me. "How old are you?" he asked, surveying me critically. "Oh, about fifty-nine," I replied. "What of it?" "Well," said he, "I have seen a cat do this but never a man." About a month since I wanted to order new eyeglasses and went to an oculist who put me thru the usual tests. He lookt at me incredulously as I read off with ease the smallest print at considerable distance. But when I told him that I was past sixty he gasped in astonishment. Friends of mine often remark that my suits fit me like gloves but they do not know that all my clothing is made to measurements which were taken nearly 35 years ago and never changed. During this same period my weight has not varied one pound.

In this connection I may tell a funny story. One evening, in the winter of 1885, Mr. Edison, Edward H. Johnson, the President of the Edison Illuminating Company, Mr. Batchellor, Manager of the works, and myself entered a little place opposite 65 Fifth Avenue where the offices of the company were located. Someone suggested guessing weights and I was induced to step on a scale. Edison felt me all over and said: "Tesla weighs 152 lbs. to an ounce," and he guest it exactly. Stript I weighed 142 lbs. and that is still my weight. I whispered to Mr. Johnson: "How is it possible that Edison could guess my weight so closely?" "Well," he said, lowering his voice. "I will tell you, confidentially, but you must not say anything. He was employed for a long time in a Chicago slaughter-house where he weighed thousands of hogs every day! That's why." My friend, the Hon. Chauncey M.

Depew, tells of an Englishman on whom he sprung one of his original anecdotes and who listened with a puzzled expression but - a year later - laughed out loud. I will frankly confess it took me longer than that to appreciate Johnson's joke.

Now, my well being is simply the result of a careful and measured mode of living and perhaps the most astonishing thing is that three times in my youth I was rendered by illness a hopeless physical wreck and given up by physicians. More than this, thru ignorance and lightheartedness, I got into all sorts of difficulties, dangers and scrapes from which I extricated myself as by enchantment. I was almost drowned a dozen times; was nearly boiled alive and just mist being cremated. I was entombed, lost and frozen. I had hair-breadth escapes from mad dogs, hogs, and other wild animals. I past thru dreadful diseases and met with all kinds of odd mishaps and that I am hale and hearty today seems like a miracle. But as I recall these incidents to my mind I feel convinced that my preservation was not altogether accidental.

An inventor's endeavor is essentially lifesaving. Whether he harnesses forces, improves devices, or provides new comforts and conveniences, he is adding to the safety of our existence. He is also better qualified than the average individual to protect himself in peril, for he is observant and resourceful. If I had no other evidence that I was, in a measure, possest of such qualities I would find it in these personal experiences. The reader will be able to judge for himself if I mention one or two instances. On one occasion, when about 14 years old, I wanted to scare some friends who were bathing with me. My plan was to dive under a long floating structure and slip out quietly at the other end. Swimming and diving came to me as naturally as to a duck and I was confident that I could perform the feat. Accordingly I plunged into the water and, when out of view, turned around and proceeded rapidly towards the opposite side. Thinking that I was safely beyond the structure, I rose to the surface but to my dismay struck a beam. Of course, I quickly dived and forged ahead with rapid strokes until my breath was beginning to give out. Rising for the second time, my head came again in contact with a beam. Now I was becoming desperate. However, summoning all my energy, I made a third frantic attempt but the result was the same. The torture of supprest breathing was getting unendurable, my brain was reeling and I felt myself sinking. At that moment, when my situation seemed absolutely hopeless, I experienced one of those flashes of light and the structure above me appeared before my vision. I either discerned or guest that there was a little space between the surface of the water and the boards resting on the beams and, with consciousness nearly gone, I floated up, prest my mouth close to the planks and managed to inhale a little air, unfortunately mingled with a spray of water which nearly choked me. Several times I repeated this procedure as in a dream until my heart, which was racing at a terrible rate, quieted down and I gained composure. After that I made a number of unsuccessful dives, having completely lost the sense of direction, but finally succeeded in getting out of the trap when my friends had already given me up and were fishing for my body.

That bathing season was spoiled for me thru recklessness but I soon forgot the lesson and only two years later I fell into a worse predicament. There was a large flour mill with a dam across the river near the city where I was studying at that time. As a rule the height of the water was only two or three inches above the dam and to swim out to it was a sport not very dangerous in which I often indulged. One day I went alone to the river to enjoy myself as usual. When I was a short distance from the masonry, however, I was horrified to observe that the water had risen and was carrying me along swiftly. I tried to get away but it was too late. Luckily, tho, I saved myself from being swept over by taking hold of the wall with both hands. The pressure against my chest was great and I was barely able to keep my head above the surface. Not a soul was in sight and my voice was lost in the roar of the fall. Slowly and gradually I became exhausted and unable to withstand the strain longer, just as I was about to let go, to be dashed against the rocks below, I saw in a flash of light a familiar diagram illustrating the hydraulic principle that the pressure of a fluid in motion is proportionate to the area exposed, and automatically I turned on my left side. As if by magic the pressure was reduced and I found it comparatively easy in that position to resist the force of the stream. But the danger still confronted me. I knew that sooner or later I would be carried down, as it was not possible for any help to reach me in time, even if I attracted attention. I am ambidextrous now but then I was lefthanded and had comparatively little strength in my right arm. For this reason I did not dare to turn on the other side to rest and nothing remained but to slowly push my body along the dam. I had to get away from the mill towards which my face was turned as the current there was much swifter and deeper. It was a long and painful ordeal and I came near to failing at its very end for I was confronted with a depression in the masonry. I managed to get over with the last ounce of my force and fell in a swoon when I reached the bank, where I was found. I had torn virtually all the skin from my left side and it took several weeks before the fever subsided and I was well. These are only two of many instances but they may be sufficient to show that had it not been for the inventor's instinct I would not have lived to tell this tale.

Interested people have often asked me how and when I began to invent. This I can only answer from my present recollection in the light of which the first attempt I recall was rather ambitious for it involved the invention of an apparatus and a method. In the former I was anticipated but the latter was original. It happened in this way. One of my playmates had come into the possession of a hook and fishing-tackle which created quite an excitement in the village, and the next morning all started out to catch frogs. I was left alone and deserted owing to a quarrel with this boy. I had never seen a real hook and pictured it as something wonderful, endowed with peculiar qualities, and was despairing not to be one of the party. Urged by necessity, I somehow got hold of a piece of soft iron wire, hammered the end to a sharp point between two stones, bent it into shape, and fastened it to a strong string. I then cut a rod, gathered some bait, and went down to the brook where there were frogs in abundance. But I could not catch any and was almost discouraged when it occurred to me to dangle the empty hook in front of a frog sitting on a stump. At first he collapsed but by and by his eyes bulged out and became bloodshot, he swelled to twice his normal size and made a vicious snap at the hook.

Immediately I pulled him up. I tried the same thing again and again and the method proved infallible. When my comrades, who in spite of their fine outfit had caught nothing, came to me they were green with envy. For a long time I kept my secret and enjoyed the monopoly but finally yielded to the spirit of Christmas. Every boy could then do the same and the following summer brought disaster to the frogs.

In my next attempt I seem to have acted under the first instinctive impulse which later dominated me - to harness the energies of nature to the service of man. I did this thru the medium of May- bugs - or June-bugs as they are called in America - which were a veritable pest in that country and sometimes broke the branches of trees by the sheer weight of their bodies. The bushes were black with them. I would attach as many as four of them to a crosspiece, ratably arranged on a thin spindle, and transmit the motion of the same to a large disc and so derive considerable "power." These creatures were remarkably efficient, for once they were started they had no sense to stop and continued whirling for hours and hours and the hotter it was the harder they worked. All went well until a strange boy came to the place. He was the son of a retired officer in the Austrian Army. That urchin ate May-bugs alive and enjoyed them as tho they were the finest blue-point oysters. That disgusting sight terminated my endeavors in this promising field and I have never since been able to touch a May-bug or any other insect for that matter.

After that, I believe, I undertook to take apart and assemble the clocks of my grandfather. In the former operation I was always successful but often failed in the latter. So it came that he brought my work to a sudden halt in a manner not too delicate and it took thirty years before I tackled another clockwork again. Shortly there after I went into the manufacture of a kind of pop-gun which comprised a hollow tube, a piston, and two plugs of hemp. When firing the gun, the piston was prest against the stomach and the tube was pushed back quickly with both hands. The air between the plugs was comprest and raised to high temperature and one of them was expelled with a loud report. The art consisted in selecting a tube of the proper taper from the hollow stalks. I did very well with that gun but my activities interfered with the window panes in our house and met with painful discouragement. If I remember rightly, I then took to carving swords from pieces of furniture which I could conveniently obtain. At that time I was under the sway of the Serbian national poetry and full of admiration for the feats of the heroes. I used to spend hours in mowing down my enemies in the form of corn-stalks which ruined the crops and retted me several spankings from my mother. Moreover these were not of the formal kind but the genuine article.

I had all this and more behind me before I was six years old and had past thru one year of elementary school in the village of Smiljan where I was born. At this juncture we moved to the little city of Gospic nearby. This change of residence was like a calamity to me. It almost broke my heart to part from our pigeons, chickens and sheep, and our magnificent flock of geese which used to rise to the clouds in the morning and return from the feeding grounds at sundown in battle formation, so perfect that it would have put a squadron of the best aviators of the present day to shame. In our new house I was but a prisoner, watching the strange people I saw thru the window blinds. My bashfulness was such that I would rather have faced a roaring lion than one of the city dudes who strolled about. But my hardest trial came on Sunday when I had to dress up and attend the service. There I meet with an accident, the mere thought of which made my blood curdle like sour milk for years afterwards. It was my second adventure in a church. Not long before I was entombed for a night in an old chapel on an inaccessible mountain which was visited only once a year. It was an awful experience, but this one was worse. There was a wealthy lady in town, a good but pompous woman, who used to come to the church gorgeously painted up and attired with an enormous train and attendants. One Sunday I had just finished ringing the bell in the belfry and rushed downstairs when this grand dame was sweeping out and I jumped on her train. It tore off with a ripping noise which sounded like a salvo of musketry fired by raw recruits. My father was livid with rage. He gave me a gentle slap on the cheek, the only corporal punishment he ever administered to me but I almost feel it now. The embarrassment and confusion that followed are indescribable. I was practically ostracised until something else happened which redeemed me in the estimation of the community.

An enterprising young merchant had organized a fire department. A new fire engine was purchased, uniforms provided and the men drilled for service and parade. The engine was, in reality, a pump to be worked by sixteen men and was beautifully painted red and black. One afternoon the official trial was prepared for and the machine was transported to the river. The entire population turned out to witness the great spectacle. When all the speeches and ceremonies were concluded, the command was given to pump, but not a drop of water came from the nozzle. The professors and experts tried in vain to locate the trouble. The fizzle was complete when I arrived at the scene. My knowledge of the mechanism was nil and I knew next to nothing of air pressure, but instinctively I felt for the suction hose in the water and found that it had collapsed. When I waded in the river and opened it up the water rushed forth and not a few Sunday clothes were spoiled. Archimedes running naked thru the streets of Syracuse and shouting Eureka at the top of his voice did not make a greater impression than myself. I was carried on the shoulders and was the hero of the day.

Upon settling in the city I began a four-years' course in the so-called Normal School preparatory to my studies at the College or Real Gymnasium. During this period my boyish efforts and exploits, as well as troubles, continued. Among other things I attained the unique distinction of champion crow catcher in the country. My method of procedure was extremely simple. I would go in the forest, hide in the bushes, and imitate the call of the bird. Usually I would get several answers and in a short while a crow would flutter down into the shrubbery near me. After that all I needed to do was to throw a piece of cardboard to distract its attention, jump up and grab it before it could extricate itself from the undergrowth. In this way I would capture as many as I desired. But on one occasion something occurred which made me respect them. I had caught a fine pair of birds and was returning home with a friend. When we left the forest, thousands of crows had gathered making a frightful racket. In a few minutes they rose in pursuit and soon enveloped us. The fun lasted until all of a sudden I received a blow on the back of my head which knocked me down. Then they attacked me viciously. I was compelled to release the two birds and was glad to join my friend who had taken refuge in a cave.

In the schoolroom there were a few mechanical models which interested me and turned my attention to water turbines. I constructed many of these and found great pleasure in operating them. How extraordinary was my life an incident may illustrate. My uncle had no use for this kind of pastime and more than once rebuked me. I was fascinated by a description of Niagara Falls I had perused, and pictured in my imagination a big wheel run by the Falls. I told my uncle that I would go to America and carry out this scheme. Thirty years later I saw my ideas carried out at Niagara and marveled at the unfathomable mystery of the mind.

I made all kinds of other contrivances and contraptions but among these the arbalists I produced were the best. My arrows, when shot, disappeared from sight and at close range traversed a plank of pine one inch thick. Thru the continuous tightening of the bows I developed skin on my stomach very much like that of a crocodile and I am often wondering whether it is due to this exercise that I am able even now to digest cobble-stones! Nor can I pass in silence my performances with the sling which would have enabled me to give a stunning exhibit at the Hippodrome. And now I will tell of one of my feats with this antique implement of war which will strain to the utmost the credulity of the reader. I was practicing while walking with my uncle along the river. The sun was setting, the trout were playful and from time to time one would shoot up into the air, its glistening body sharply defined against a projecting rock beyond. Of course any boy might have hit a fish under these propitious conditions but I undertook a much more difficult task and I foretold to my uncle, to the minutest detail, what I intended doing. I was to hurl a stone to meet the fish, press its body against the rock, and cut it in two. It was no sooner said than done. My uncle looked at me almost scared out of his wits and exclaimed "Vade retro Satanas!" and it was a few days before he spoke to me again. Other records, how ever great, will be eclipsed but I feel that I could peacefully rest on my laurels for a thousand years.

III. My Later Endeavors

The Discovery of the Rotating Magnetic Field

At the age of ten I entered the Real Gymnasium which was a new and fairly well equipt institution. In the department of physics were various models of classical scientific apparatus, electrical and mechanical. The demonstrations and experiments performed from time to time by the instructors fascinated me and were undoubtedly a powerful incentive to invention. I was also passionately fond of mathematical studies and often won the professor's praise for rapid calculation. This was due to my acquired facility of visualizing the figures and performing the operations, not in the usual intuitive manner, but as in actual life. Up to a certain degree of complexity it was absolutely the same to me whether I wrote the symbols on the board or conjured them before my mental vision. But freehand drawing, to which many hours of the course were devoted, was an annoyance I could not endure. This was rather remarkable as most of the members of the family excelled in it. Perhaps my aversion was simply due to the predilection I found in undisturbed thought. Had it not been for a few exceptionally stupid boys, who could not do anything at all, my record would have been the worst. It was a serious handicap as under the then existing educational regime, drawing being obligatory, this deficiency threatened to spoil my whole career and my father had considerable trouble in railroading me from one class to another.

In the second year at that institution I became obsessed with the idea of producing continuous motion thru steady air pressure. The pump incident, of which I have told, had set afire my youthful imagination and imprest me with the boundless abilities of a vacuum. I grew frantic in my desire to harness this inexhaustible energy but for a long time I was groping in the dark. Finally, however, my endeavors crystallized in an invention which was to enable me to achieve what no other mortal ever attempted.

Imagine a cylinder freely rotatable on two bearings and partly surrounded by a rectangular trough which fits it perfectly. The open side of the trough is closed by a partition so that the cylindrical segment within the enclosure divides the latter into two compartments entirely separated from each other by air-tight sliding joints. One of these compartments being sealed and once for all exhausted, the other remaining open, a perpetual rotation of the cylinder would result, at least, I thought so. A wooden model was constructed and fitted with infinite care and when I applied the pump on one side and actually observed that there was a tendency to turning, I was delirious with joy. Mechanical flight was the one thing I wanted to accomplish altho still under the discouraging recollection of a bad fall I sustained by jumping with an umbrella from the top of a building. Every day I used to transport myself thru the air to distant regions but could not understand just how I managed to do it. Now I had something concrete — a flying machine with nothing more than a rotating shaft, flapping wings, and — a vacuum of unlimited power! From that time on I made my daily aerial excursions in a vehicle of comfort and luxury as might have befitted King Solomon. It took years before I understood that the atmospheric pressure acted at right angles to the surface of the cylinder and that the slight rotary effort I observed was due to a leak. Tho this knowledge came gradually it gave me a painful shock.

I had hardly completed my course at the Real Gymnasium when I was prostrated with a dangerous illness or rather, a score of them, and my condition became so desperate that I was given up by physicians. During this period I was permitted to read constantly, obtaining books from the Public Library which had been neglected and entrusted to me for classification of the works and preparation of the catalogues. One day I was handed a few volumes of new literature unlike anything I had ever read before and so captivating as to make me utterly forget my hopeless state. They were the earlier works of Mark Twain and to them might have been due the miraculous recovery which followed. Twenty-five years later, when I met Mr. Clemens and we formed a friendship between us, I told him of the experience and was amazed to see that great man of laughter burst into tears.

My studies were continued at the higher Real Gymnasium in Carlstadt, Croatia, where one of my aunts resided. She was a distinguished lady, the wife of a Colonel who was an old war-horse having participated in many battles. I never can forget the three years I past at their home. No fortress in time of war was under a more rigid discipline. I was fed like a canary bird. All the meals were of the highest quality and deliciously prepared but short in quantity by a thousand percent. The slices of ham cut by my aunt were like tissue paper. When the Colonel would put something substantial on my plate she would snatch it away and say excitedly to him: "Be careful, Niko is very delicate." I had a voracious appetite and suffered like Tantalus. But I lived in an atmosphere of refinement and artistic taste quite unusual for those times and conditions. The land was low and marshy and malaria fever never left me while there despite of the enormous amounts of quinin I consumed. Occasionally the river would rise and drive an army of rats into the buildings, devouring everything even to the bundles of the fierce paprika. These pests were to me a welcome diversion. I thinned their ranks by all sorts of means, which won me the unenviable distinction of ratcatcher in the community. At last, however, my course was completed, the misery ended, and I obtained the certificate of maturity which brought me to the cross-roads.

During all those years my parents never wavered in their resolve to make me embrace the clergy, the mere thought of which filled me with dread. I had become intensely interested in electricity under the stimulating influence of my Professor of Physics, who was an ingenious man and often demonstrated the principles by apparatus of his own invention. Among these I recall a device in the shape of a freely rotatable bulb, with tinfoil coatings, which was made to spin rapidly when connected to a static machine. It is impossible for me to convey an adequate idea of the intensity of feeling I experienced in witnessing his exhibitions of these mysterious phenomena. Every impression produced a thousand echoes in my mind. I wanted to know more of this wonderful force; I longed for experiment and investigation and resigned myself to the inevitable with aching heart.

Just as I was making ready for the long journey home I received word that my father wished me to go on a shooting expedition. It was a strange request as he had been always strenuously opposed to this kind of sport. But a few days later I learned that the cholera was raging in that district and, taking advantage of an opportunity, I returned to Gospic in disregard of my parents' wishes. It is incredible how absolutely ignorant people were as to the causes of this scourge which visited the country in intervals of from fifteen to twenty years. They thought that the deadly agents were transmitted thru the air and filled it with pungent odors and smoke. In the meantime they drank the infected water and died in heaps. I contracted the awful disease on the very day of my arrival and altho surviving the crisis, I was confined to bed for nine months with scarcely any ability to move. My energy was completely exhausted and for the second time I found myself at death's door. In one of the sinking spells which was thought to be the last, my father rushed into the room. I still see his pallid face as he tried to cheer me in tones belying his assurance. "Perhaps," I said, "I may get well if you will let me study engineering." "You will go to the best technical institution in the world," he solemnly replied, and I knew that he meant it. A heavy weight was lifted from my mind but the relief would have come too late had it not been for a marvelous cure brought about thru a bitter decoction of a peculiar bean. I came to life like another Lazarus to the utter amazement of everybody.

My father insisted that I spend a year in healthful physical outdoor exercises to which I reluctantly consented. For most of this term I roamed in the mountains, loaded with a hunter's outfit and a bundle of books, and this contact with nature made me stronger in body as well as in mind. I thought and planned, and conceived many ideas almost as a rule delusive. The vision was clear enough but the knowledge of principles was very limited. In one of my inventions I proposed to convey letters and packages across the seas, thru a submarine tube, in spherical containers of sufficient strength to resist the hydraulic pressure. The pumping plant, intended to force the water thru the tube, was accurately figured and designed and all other particulars carefully worked out. Only one trifling detail, of no consequence, was lightly dismist. I assumed an arbitrary velocity of the water and, what is more, took pleasure in making it high, thus arriving at a stupendous performance supported by faultless calculations. Subsequent reflections, however, on the resistance of pipes to fluid flow determined me to make this invention public property.

Another one of my projects was to construct a ring around the equator which would, of course, float freely and could be arrested in its spinning motion by reactionary forces, thus enabling travel at a rate of about one thousand miles an hour, impracticable by rail. The reader will smile. The plan was difficult of execution, I will admit, but not nearly so bad as that of a well-known New York professor, who wanted to pump the air from the torrid to the temperate zones, entirely forgetful of the fact that the Lord had provided a gigantic machine for this very purpose.

Still another scheme, far more important and attractive, was to derive power from the rotational energy of terrestrial bodies. I had discovered that objects on the earth's surface, owing to the diurnal rotation of the globe, are carried by the same alternately in and against the direction of translatory movement. From this results a great change in momentum which could be utilized in the simplest imaginable manner to furnish motive effort in any habitable region of the world. I cannot find words to describe my disappointment when later I realized that I was in the predicament of Archimedes, who vainly sought for a fixt point in the universe.

At the termination of my vacation I was sent to the Polytechnic School in Gratz, Styria, which my father had chosen as one of the oldest and best reputed institutions. That was the moment I had eagerly awaited and I began my studies under good auspices and firmly resolved to succeed. My previous training was above the average, due to my father's teaching and opportunities afforded. I had acquired the knowledge of a number of languages and waded thru the books of several libraries, picking up information more or less useful. Then again, for the first time, I could choose my subjects as I liked, and free-hand drawing was to bother me no more.

I had made up my mind to give my parents a surprise, and during the whole first year I regularly started my work at three o'clock in the morning and continued until eleven at night, no Sundays or holidays excepted. As most of my fellow -students took thinks easily, naturally enough I eclipsed all records. In the course of that year I past thru nine exams and the professors thought I deserved more than the highest qualifications. Armed with their flattering certificates, I went home for a short rest, expecting a triumph, and was mortified when my father made light of these hard won honors. That almost killed my ambition; but later, after he had died, I was pained to find a package of letters which the professors had written him to the effect that unless he took me away from the Institution I would be killed thru overwork.

Thereafter I devoted myself chiefly to physics, mechanics and mathematical studies, spending the hours of leisure in the libraries. I had a veritable mania for finishing whatever I began, which often got me into difficulties. On one occasion I started to read the works of Voltaire when I learned, to my dismay, that there were close on one hundred large volumes in small print which that monster had written while drinking seventy-two cups of black coffee per diem. It had to be done, but when I laid aside the last book I was very glad, and said, "Never more!"

My first year's showing had won me the appreciation and friendship of several professors. Among these were Prof. Rogner, who was teaching arithmetical subjects and geometry; Prof. Poeschl, who held the chair of theoretical and experimental physics, and Dr. Alle, who taught integral calculus and specialized in differential equations. This scientist was the most brilliant lecturer to whom I ever listened. He took a special interest in my progress and would frequently remain for an hour or two in the lecture room, giving me problems to solve, in which I delighted. To him I explained a flying machine I had conceived, not an illusionary invention, but one based on sound, scientific principles, which has become realizable thru my turbine and will soon be given to the world. Both Professors Rogner and Poeschl were curious men. The former had peculiar ways of expressing himself and whenever he did so there was a riot, followed by a long and embarrassing pause. Prof. Poeschl was a methodical and thoroly grounded German. He had enormous feet and hands like the paws of a bear, but all of his experiments were skillfully performed with lock-like precision and without a miss.

It was in the second year of my studies that we received a Gramme dynamo from Paris, having the horseshoe form of a laminated field magnet, and a wire-wound armature with a commutator. It was connected up and various effects of the currents were shown. While Prof. Poeschl was making demonstrations, running the machine as a motor, the brushes gave trouble, sparking badly, and I observed that it might be possible to operate a motor without these appliances. But he declared that it could not be done and did me the honor of delivering a lecture on the subject, at the conclusion of which he remarked: "Mr. Tesla may accomplish great things, but he certainly never will do this. It would be equivalent to converting a steadily pulling force, like that of gravity, into a rotary effort. It is a perpetual motion scheme, an impossible idea." But instinct is something which transcends knowledge. We have, undoubtedly, certain finer fibers that enable us to perceive truths when logical deduction, or any other willful effort of the brain, is futile. For a time I wavered, imprest by the professor's authority, but soon became convinced I was right and undertook the task with all the fire and boundless confidence of youth.

I started by first picturing in my mind a direct-current machine, running it and following the changing flow of the currents in the armature. Then I would imagine an alternator and investigate the processes taking place in a similar manner. Next I would visualize systems comprising motors and generators and operate them in various ways. The images I saw were to me perfectly real and tangible. All my remaining term in Gratz was passed in intense but fruitless efforts of this kind, and I almost came to the conclusion that the problem was insolvable.

In 1880 I went to Prague, Bohemia, carrying out my father's wish to complete my education at the University there. It was in that city that I made a decided advance, which consisted in detaching the commutator from the machine and studying the phenomena in this new aspect, but still without result. In the year following there was a sudden change in my views of life. I realized that my parents had been making too great sacrifices on my account and resolved to relieve them of the burden. The wave of the American telephone had just reached the European continent and the system was to be installed in Budapest, Hungary. It appeared an ideal opportunity, all the more as a friend of our family was at the head of the enterprise. It was here that I suffered the complete breakdown of the nerves to which I have referred.

What I experienced during the period of that illness surpasses all belief. My sight and hearing were always extraordinary. I could clearly discern objects in the distance when others saw no trace of them. Several times in my boyhood I saved the houses of our neighbors from fire by hearing the faint crackling sounds which did not disturb their sleep, and calling for help.

In 1899, when I was past forty and carrying on my experiments in Colorado, I could hear very distinctly thunderclaps at a distance of 550 miles. The limit of audition for my young assistants was scarcely more than 150 miles. My ear was thus over thirteen times more sensitive. Yet at that time I was, so to speak, stone deaf in comparison with the acuteness of my hearing while under the nervous strain. In Budapest I could hear the ticking of a watch with three rooms between me and the time-piece. A fly alighting on a table in the room would cause a dull thud in my ear. A carriage passing at a distance of a few miles fairly shook my whole body. The whistle of a locomotive twenty or thirty miles away made the bench or chair on which I sat vibrate so strongly that the pain was unbearable. The ground under my feet trembled continuously. I had to support my bed on rubber cushions to get any rest at all. The roaring noises from near and far often produced the effect of spoken words which would have frightened me had I not been able to resolve them into their accidental components. The sun's rays, when periodically intercepted, would cause blows of such force on my brain that they would stun me. I had to summon all my will power to pass under a bridge or other structure as I experienced a crushing pressure on the skull. In the dark I had the sense of a bat and could detect the presence of an object at a distance of twelve feet by a peculiar creepy sensation on the forehead. My pulse varied from a few to two hundred and sixty beats and all the tissues of the body quivered with twitchings and tremors which was perhaps the hardest to bear. A renowned physician who gave me daily large doses of Bromide of Potassium pronounced my malady unique and incurable.

It is my eternal regret that I was not under the observation of experts in physiology and psychology at that time. I clung desperately to life, but never expected to recover. Can anyone believe that so hopeless a physical wreck could ever be transformed into a man of astonishing strength and tenacity, able to work thirty-eight years almost without a day's interruption, and find himself still strong and fresh in body and mind? Such is my case. A powerful desire to live and to continue the work, and the assistance of a devoted friend and athlete accomplished the wonder. My health returned and with it the vigor of mind. In attacking the problem again I almost regretted that the struggle was soon to end. I had so much energy to spare. When I undertook the task it was not with a resolve such as men often make. With me it was a sacred vow, a question of life and death. I knew that I would perish if I failed. Now I felt that the battle was won. Back in the deep recesses of the brain was the solution, but I could not yet give it outward expression. One afternoon, which is ever present in my recollection, I was enjoying a walk with my friend in the City Park and reciting poetry. At that age I knew entire books by heart, word for word. One of these was Goethe's "Faust." The sun was just setting and reminded me of the glorious passage:

"Sie ruckt und weicht, der Tag ist uberlebt, Dort eilt sie hin und fordert neues Leben. Oh, dass kein Flugel mich vom Boden hebt Ihr nach und immer nach zu streben!

Ein schoner Traum indessen sie entweicht, Ach, zu des Geistes Flugeln wird so leicht Kein korperlicher Flugel sich gesellen!"

[The glow retreats, done is the day of toil; It yonder hastes, new fields of life exploring; Ah, that no wing can lift me from the soil Upon its track to follow, follow soaring!

A glorious dream! though now the glories fade. Alas! the wings that lift the mind no aid Of wings to lift the body can bequeath me.]

As I uttered these inspiring words the idea came like a flash of lightning and in an instant the truth was revealed. I drew with a stick on the sand the diagrams shown six years later in my address before the American Institute of Electrical Engineers, and my companion understood them perfectly. The images I saw were wonderfully sharp and clear and had the solidity of metal and stone, so much so that I told him: "See my motor here; watch me reverse it." I cannot begin to describe my emotions. Pygmalion seeing his statue come to life could not have been more deeply moved. A thousand secrets of nature which I might have stumbled upon accidentally I would have given for that one which I had wrested from her against all odds and at the peril of my existence.

IV. The Discovery of the Tesla Coil and Transformer

For a while I gave myself up entirely to the intense enjoyment of picturing machines and devising new forms. It was a mental state of happiness about as complete as I have ever known in life. Ideas came in an uninterrupted stream and the only difficulty I had was to hold them fast. The pieces of apparatus I conceived were to me absolutely real and tangible in every detail, even to the minute marks and signs of wear. I delighted in imagining the motors constantly running, for in this way they presented to mind's eye a more fascinating sight. When natural inclination develops into a passionate desire, one advances towards his goal in seven-league boots. In less than two months I evolved virtually all the types of motors and modifications of the system which are now identified with my name. It was, perhaps, providential that the necessities of existence commanded a temporary halt to this consuming activity of the mind. I came to Budapest prompted by a premature report concerning the telephone enterprise and, as irony of fate willed it, I had to accept a position as draftsman in the Central Telegraph Office of the Hungarian Government at a salary which I deem it my privilege not to disclose! Fortunately, I soon won the interest of the Inspector-in-Chief and was thereafter employed on calculations, designs and estimates in connection with new installations, until the Telephone Exchange was started, when I took charge of the same. The knowledge and practical experience I gained in the course of this work was most valuable and the employment gave me ample opportunities for the exercise of my inventive faculties. I made several improvements in the Central Station apparatus and perfected a telephone repeater or amplifier which was never patented or publicly described but would be creditable to me even today. In recognition of my efficient assistance the organizer of the undertaking, Mr. Puskas, upon disposing of his business in Budapest, offered me a position in Paris which I gladly accepted.

I never can forget the deep impression that magic city produced on my mind. For several days after my arrival I roamed thru the streets in utter bewilderment of the new spectacle. The attractions were many and irresistible, but, alas, the income was spent as soon as received. When Mr. Puskas asked me how I was getting along in the new sphere, I described the situation accurately in the statement that "the last twenty-nine days of the month are the toughest!" I led a rather strenuous life in what would now be termed "Rooseveltian fashion." Every morning, regardless of weather, I would go from the Boulevard St. Marcel, where I resided, to a bathing house on the Seine, plunge into the water, loop the circuit twenty-seven times and then walk an hour to reach Ivry, where the Company's factory was located. There I would have a woodchopper's breakfast at half-past seven o'clock and then eagerly await the lunch hour, in the meanwhile cracking hard nuts for the Manager of the Works, Mr. Charles Batchellor, who was an intimate friend and assistant of Edison. Here I was thrown in contact with a few Americans who fairly fell in love with me because of my proficiency in billiards. To these men I explained my invention and one of them, Mr. D. Cunningham, Foreman of the Mechanical Department, offered to form a stock company. The proposal seemed to me comical in the extreme. I did not have the faintest conception of what that meant except that it was an American way of doing things. Nothing came of it, however, and during the next few months I had to travel from one to another place in France and Germany to cure the ills of the power plants. On my return to Paris I submitted to one of the administrators of the

42

Company, Mr. Rau, a plan for improving their dynamos and was given an opportunity. My success was complete and the delighted directors accorded me the privilege of developing automatic regulators which were much desired. Shortly after there was some trouble with the lighting plant which had been installed at the new railroad station in Strassburg, Alsace. The wiring was defective and on the occasion of the opening ceremonies a large part of a wall was blown out thru a short-circuit right in the presence of old Emperor William I. The German Government refused to take the plant and the French Company was facing a serious loss. On account of my knowledge of the German language and past experience, I was entrusted with the difficult task of straightening out matters and early in 1883 I went to Strassburg on that mission.

Some of the incidents in that city have left an indelible record on my memory. By a curious coincidence, a number of men who subsequently achieved fame, lived there about that time. In later life I used to say, "There were bacteria of greatness in that old town. Others caught the disease but I escaped!" The practical work, correspondence, and conferences with officials kept me preoccupied day and night, but, as soon as I was able to manage I undertook the construction of a simple motor in a mechanical shop opposite the railroad station, having brought with me from Paris some material for that purpose. The consummation of the experiment was, however, delayed until the summer of that year when I finally had the satisfaction of seeing rotation effected by alternating currents of different phase, and without sliding contacts or commutator, as I had conceived a year before. It was an exquisite pleasure but not to compare with the delirium of joy following the first revelation.

Among my new friends was the former Mayor of the city, Mr. Bauzin, whom I had already in a measure acquainted with this and other inventions of mine and whose support I endeavored to enlist. He was sincerely devoted to me and put my project before several wealthy persons but, to my mortification, found no response. He wanted to help me in every possible way and the approach of the first of July, 1919, happens to remind me of a form of "assistance" I received from that charming man, which was not financial but none the less appreciated. In 1870, when the Germans invaded the country, Mr. Bauzin had buried a good sized allotment of St. Estephe of 1801 and he came to the conclusion that he knew no worthier person than myself to consume that precious beverage. This, I may say, is one of the unforgettable incidents to which I have referred. My friend urged me to return to Paris as soon as possible and seek support there. This I was anxious to do but my work and negotiations were protracted owing to all sorts of petty obstacles I encountered so that at times the situation seemed hopeless.

Just to give an idea of German thoroness and "efficiency," I may mention here a rather funny experience. An incandescent lamp of 16 c.p. was to be placed in a hallway and upon selecting the proper location I ordered the monteur to run the wires. After working for a while he concluded that the engineer had to be consulted and this was done. The latter made several objections but ultimately agreed that the lamp should be placed two inches from the spot I had assigned, whereupon the work proceeded. Then the engineer became worried and told me that Inspector Averdeck should be notified. That important person called, investigated, debated, and decided that the lamp should be shifted back two inches, which was the place I had marked. It was not long, however, before Averdeck got cold feet himself and advised me that he had informed Ober- Inspector Hieronimus of the matter and that I should await his decision. It was several days before the Ober- Inspector was able to free himself of other pressing duties but at last he arrived and a two-hour debate followed, when he decided to move the lamp two inches farther. My hopes that this was the final act were shattered when the Ober-lnspector returned and said to me: "Regierungsrath Funke is so particular that I would not dare to give an order for placing this lamp without his explicit approval." Accordingly arrangements for a visit from that great man were made. We started cleaning up and polishing early in the morning. Everybody brushed up, I put on my gloves and when Funke came with his retinue he was ceremoniously received. After two hours' deliberation he suddenly exclaimed: "I must be going," and pointing to a place on the ceiling, he ordered me to put the lamp there. It was the exact spot which I had originally chosen,

So it went day after day with variations, but I was determined to achieve at whatever cost and in the end my efforts were rewarded. By the spring of 1884 all the differences were adjusted, the plant formally accepted, and I returned to Paris with pleasing anticipations. One of the administrators had promised me a liberal compensation in case I succeeded, as well as a fair consideration of the improvements I had made in their dynamos and I hoped to realize a substantial sum. There were three administrators whom I shall designate as A, B and C for convenience. When I called on A he told me that B had the say. This gentleman thought that only C could decide and the latter was quite sure that A alone had the power to act. After several laps of this circulus vivios it dawned upon me that my reward was a castle in Spain. The utter failure of my attempts to raise capital for development was another disappointment and when Mr. Batchellor prest me to go to America with a view of redesigning the Edison machines, I determined to try my fortunes in the Land of Golden Promise. But the chance was nearly mist. I liquefied my modest assets, secured accommodations and found myself at the railroad station as the train was pulling out. At that moment I discovered that my money and tickets were gone. What to do was the question. Hercules had plenty of time to deliberate but I had to decide while running alongside the train with opposite feelings surging in my brain like condenser oscillations. Resolve, helped by dexterity, won out in the nick of time and upon passing thru the usual experiences, as trivial as unpleasant, I managed to embark for New York with the remnants of my belongings, some poems and articles I had written, and a package of calculations relating to solutions of an unsolvable integral and to my flying machine. During the voyage I sat most of the time at the stern of the ship watching for an opportunity to save somebody from a watery grave, without the slightest thought of danger. Later when I had absorbed some of the practical American sense I shivered at the recollection and marvelled at my former folly.

I wish that I could put in words my first impressions of this country. In the Arabian Tales I read how genii transported people into a land of dreams to live thru delightful adventures. My case was just the reverse. The genii had carried me from a world of dreams into one of realities. What I had left was beautiful, artistic and fascinating in every way; what I saw here was machined, rough and unattractive. A burly policeman was twirling his stick which looked to me as big as a log. I approached him politely with the request to direct me. "Six blocks down, then to the left," he said, with murder in his eyes. "Is this America?" I asked myself in painful surprise. "It is a century behind Europe in civilization." When I went abroad in 1889 - five years having elapsed since my arrival here - 1 became convinced that it was more than one hundred years AHEAD of Europe and nothing has happened to this day to change my opinion.

The meeting with Edison was a memorable event in my life. I was amazed at this wonderful man who, without early advantages and scientific training, had accomplished so much. I had studied a dozen languages, delved in literature and art, and had spent my best years in libraries reading all sorts of stuff that fell into my hands, from Newton's "Principia" to the novels of Paul de Kock, and felt that most of my life had been squandered. But it did not take long before I recognized that it was the best thing I could have done. Within a few weeks I had won Edison's confidence and it came about in this way.

The S.S. Oregon, the fastest passenger steamer at that time, had both of its lighting machines disabled and its sailing was delayed. As the superstructure had been built after their installation it was impossible to remove them from the hold. The predicament was a serious one and Edison was much annoyed. In the evening I took the necessary instruments with me and went aboard the vessel where I stayed for the night. The dynamos were in bad condition, having several short-circuits and breaks, but with the assistance of the crew I succeeded in putting them in good shape. At five o'clock in the morning, when passing along Fifth Avenue on my way to the shop, I met Edison with Batchellor and a few others as they were returning home to retire. "Here is our Parisian running around at night," he said. When I told him that I was coming from the Oregon and had repaired both machines, he looked at me in silence and walked away without another word. But when he had gone some distance I heard him remark: "Batchellor, this is a d-n good man," and from that time on I had full freedom in directing the work. For nearly a year my regular hours were from 10.30 A.M. until 5 o'clock the next morning without a day's exception. Edison said to me: "I have had many hard-working assistants but you take the cake." During this period I designed twenty-four different types of standard machines with short cores and of uniform pattern which replaced the old ones. The Manager had promised me fifty thousand dollars on the completion of this task but it turned out to be a practical joke. This gave me a painful shock and I resigned my position.

Immediately thereafter some people approached me with the proposal of forming an arc light company under my name, to which I agreed. Here finally was an opportunity to develop the motor, but when I broached the subject to my new associates they said: "No, we want the arc lamp. We don't care for this alternating current of yours." In 1886 my system of arc lighting was perfected and adopted for factory and municipal lighting, and I was free, but with no other possession than a beautifully engraved certificate of stock of hypothetical value. Then followed a period of struggle in the new medium for which I was not fitted, but the reward came in the end and in April, 1887, the Tesla Electric Company was organized, providing a laboratory and facilities. The motors I built there were exactly as I had imagined them. I made no attempt to improve the design, but merely reproduced the pictures as they appeared to my vision and the operation was always as I expected.

In the early part of 1888 an arrangement was made with the Westinghouse Company for the manufacture of the motors on a large scale. But great difficulties had still to be overcome. My system was based on the use of low frequency currents and the Westinghouse experts had adopted 133 cycles with the object of securing advantages in the transformation. They did not want to depart from their standard forms of apparatus and my efforts had to be concentrated upon adapting the motor to these conditions. Another necessity was to produce a motor capable of running efficiently at this frequency on two wires which was not easy of accomplishment.

At the close of 1889, however, my services in Pittsburg being no longer essential, I returned to New York and resumed experimental work in a laboratory on Grand Street, where I began immediately the design of high frequency machines. The problems of construction in this unexplored field were novel and quite peculiar and I encountered many difficulties. I rejected the inductor type, fearing that it might not yield perfect sine waves which were so important to resonant action. Had it not been for this I could have saved myself a great deal of labor. Another discouraging feature of the high frequency alternator seemed to be the inconstancy of speed which threatened to impose serious limitations to its use. I had already noted in my demonstrations before the American Institution of Electrical Engineers that several times the tune was lost, necessitating readjustment, and did not yet foresee, what I discovered long afterwards, a means of operating a machine of this kind at a speed constant to such a degree as not to vary more than a small fraction of one revolution between the extremes of load.

From many other considerations it appeared desirable to invent a simpler device for the production of electric oscillations. In 1856 Lord Kelvin had exposed the theory of the condenser discharge, but no practical application of that important knowledge was made. I saw the possibilities and undertook the development of induction apparatus on this principle. My progress was so rapid as to enable me to exhibit at my lecture in 1891 a coil giving sparks of five inches. On that occasion I frankly told the engineers of a defect involved in the transformation by the new method, namely, the loss in the spark gap. Subsequent investigation showed that no matter what medium is employed, be it air, hydrogen, mercury vapor, oil or a stream of electrons, the efficiency is the same. It is a law very much like that governing the conversion of mechanical energy. We may drop a weight from a certain height vertically down or carry it to the

49

lower level along any devious path, it is immaterial insofar as the amount of work is concerned. Fortunately however, this drawback is not fatal as by proper proportioning of the resonant circuits an efficiency of 85 per cent is attainable. Since my early announcement of the invention it has come into universal use and wrought a revolution in many departments. But a still greater future awaits it. When in 1900 I obtained powerful discharges of 100 feet and flashed a current around the globe, I was reminded of the first tiny spark I observed in my Grand Street laboratory and was thrilled by sensations akin to those I felt when I discovered the rotating magnetic field.

V.   The Magnifying Transmitter

As I review the events of my past life I realize how subtle are the influences that shape our destinies. An incident of my youth may serve to illustrate. One winter's day I managed to climb a steep mountain, in company with other boys. The snow was quite deep and a warm southerly wind made it just suitable for our purpose. We amused ourselves by throwing balls which would roll down a certain distance, gathering more or less snow, and we tried to outdo one another in this exciting sport. Suddenly a ball was seen to go beyond the limit, swelling to enormous proportions until it became as big as a house and plunged thundering into the valley below with a force that made the ground tremble. I looked on spellbound, incapable of understanding what had happened. For weeks afterward the picture of the avalanche was before my eyes and I wondered how anything so small could grow to such an immense size. Ever since that time the magnification of feeble actions fascinated me, and when, years later, I took up the experimental study of mechanical and electrical resonance, I was keenly interested from the very start. Possibly, had it not been for that early powerful impression, I might not have followed up the little spark I obtained with my coil and never developed my best invention, the true history of which I'll tell here for the first time.

"Lionhunters" have often asked me which of my discoveries I prize most. This depends on the point of view. Not a few technical men, very able in their special departments, but dominated by a pedantic spirit and nearsighted, have asserted that excepting the induction motor I have given to the world little of practical use. This is a grievous mistake. A new idea must not be judged by its immediate results. My alternating system of

51

power transmission came at a psychological moment, as a long-sought answer to pressing industrial questions, and altho considerable resistance had to be overcome and opposing interests reconciled, as usual, the commercial introduction could not be long delayed. Now, compare this situation with that confronting my turbine, for example. One should think that so simple and beautiful an invention, possessing many features of an ideal motor, should be adopted at once and, undoubtedly, it would under similar conditions. But the prospective effect of the rotating field was not to render worthless existing machinery; on the contrary, it was to give it additional value. The system lent itself to new enterprise as well as to improvement of the old. My turbine is an advance of a character entirely different. It is a radical departure in the sense that its success would mean the abandonment of the antiquated types of prime movers on which billions of dollars have been spent. Under such circumstances the progress must needs be slow and perhaps the greatest impediment is encountered in the prejudicial opinions created in the minds of experts by organized opposition.

Only the other day I had a disheartening experience when I met my friend and former assistant, Charles F. Scott, now professor of Electrical Engineering at Yale. I had not seen him for a long time and was glad to have an opportunity for a little chat at my office. Our conversation naturally enough drifted on my turbine and I became heated to a high degree. "Scott," I exclaimed, carried away by the vision of a glorious future, "my turbine will scrap all the heat-engines in the world." Scott stroked his chin and looked away thoughtfully, as though making a mental calculation. "That will make quite a pile of scrap," he said, and left without another word!

These and other inventions of mine, however, were nothing more than steps forward in certain directions. In evolving them

I simply followed the inborn sense to improve the present devices without any special thought of our far more imperative necessities. The "Magnifying Transmitter" was the product of labors extending through years, having for their chief object the solution of problems which are infinitely more important to mankind than mere industrial development.

If my memory serves me right, it was in November, 1890, that I performed a laboratory experiment which was one of the most extraordinary and spectacular ever recorded in the annals of science. In investigating the behaviour of high frequency currents I had satisfied myself that an electric field of sufficient intensity could be produced in a room to light up electrodeless vacuum tubes. Accordingly, a transformer was built to test the theory and the first trial proved a marvelous success. It is difficult to appreciate what those strange phenomena meant at that time. We crave for new sensations but soon become indifferent to them. The wonders of yesterday are today common occurrences. When my tubes were first publicly exhibited they were viewed with amazement impossible to describe. From all parts of the world I received urgent invitations and numerous honors and other flattering inducements were offered to me, which I declined.

But in 1892 the demands became irresistible and I went to London where I delivered a lecture before the Institution of Electrical Engineers. It had been my intention to leave immediately for Paris in compliance with a similar obligation, but Sir James Dewar insisted on my appearing before the Royal Institution. I was a man of firm resolve but succumbed easily to the forceful arguments of the great Scotsman. He pushed me into a chair and poured out half a glass of a wonderful brown fluid which sparkled in all sorts of iridescent colors and tasted like nectar. "Now," said he. "you are sitting in Faraday's chair and you are enjoying whiskey he used to drink." In both aspects it was an enviable experience. The next evening I gave a demonstration before that Institution, at the termination of which Lord Rayleigh addressed the audience and his generous words gave me the first start in these endeavors. I fled from London and later from Paris to escape favors showered upon me, and journeyed to my home where I passed through a most painful ordeal and illness. Upon regaining my health I began to formulate plans for the resumption of work in America. Up to that time I never realized that I possessed any particular gift of discovery but Lord Rayleigh, whom I always considered as an ideal man of science, had said so and if that was the case I felt that I should concentrate on some big idea.

One day, as I was roaming in the mountains, I sought shelter from an approaching storm. The sky became overhung with heavy clouds but somehow the rain was delayed until, all of a sudden, there was a lightning flash and a few moments after a deluge. This observation set me thinking. It was manifest that the two phenomena were closely related, as cause and effect, and a little reflection led me to the conclusion that the electrical energy involved in the precipitation of the water was inconsiderable, the function of lightning being much like that of a sensitive trigger.

Here was a stupendous possibility of achievement. If we could produce electric effects of the required quality, this whole planet and the conditions of existence on it could be transformed. The sun raises the water of the oceans and winds drive it to distant regions where it remains in a state of most delicate balance. If it were in our power to upset it when and wherever desired, this mighty life-sustaining stream could be at will controlled. We could irrigate arid deserts, create lakes and rivers and provide motive power in unlimited amounts. This would be the most efficient way of harnessing the sun to the uses of man. The consummation depended on our ability to develop electric forces of the order of those in nature. It seemed a hopeless undertaking, but I made up my mind to try it and immediately on my return to the United States, in the Summer of 1892, work was begun which was to me all the more attractive, because a means of the same kind was necessary for the successful transmission of energy without wires.

The first gratifying result was obtained in the spring of the succeeding year when I reached tensions of about 1 ,000,000 volts with my conical coil. That was not much in the light of the present art, but it was then considered a feat. Steady progress was made until the destruction of my laboratory by fire in 1895, as may be judged from an article by T. C. Martin which appeared in the April number of the Century Magazine. This calamity set me back in many ways and most of that year had to be devoted to planning and reconstruction. However, as soon as circumstances permitted, I returned to the task.

Although I knew that higher electro-motive forces were attainable with apparatus of larger dimensions, I had an instinctive perception that the object could be accomplished by the proper design of a comparatively small and compact transformer. In carrying on tests with a secondary in the form of a flat spiral, as illustrated in my patents, the absence of streamers surprised me, and it was not long before I discovered that this was due to the position of the turns and their mutual action. Profiting from this observation I resorted to the use of a high tension conductor with turns of considerable diameter sufficiently separated to keep down the distributed capacity, while at the same time preventing undue accumulation of the charge at any point. The application of this principle enabled me to produce pressures of 4,000,000 volts, which was about the limit obtainable in my new laboratory at Houston Street, as the discharges extended through a distance of 16 feet. A photograph of this transmitter was published in the Electrical Review of November, 1898.

In order to advance further along this line I had to go into the open, and in the spring of 1899, having completed preparations for the erection of a wireless plant, I went to Colorado where I remained for more than one year. Here I introduced other improvements and refinements which made it possible to generate currents of any tension that may be desired. Those who are interested will find some information in regard to the experiments I conducted there in my article, "The Problem of Increasing Human Energy" in the Century Magazine of June, 1900, to which I have referred on a previous occasion.

I have been asked by the ELECTRICAL EXPERIMENTER to be quite explicit on this subject so that my young friends among the readers of the magazine will clearly understand the construction and operation of my "Magnifying Transmitter" and the purposes for which it is intended. Well, then, in the first place, it is a resonant transformer with a secondary in which the parts, charged to a high potential, are of considerable area and arranged in space along ideal enveloping surfaces of very large radii of curvature, and at proper distances from one another thereby insuring a small electric surface density everywhere so that no leak can occur even if the conductor is bare. It is suitable for any frequency, from a few to many thousands of cycles per second, and can be used in the production of currents of tremendous volume and moderate pressure, or of smaller amperage and immense electromotive force. The maximum electric tension is merely dependent on the curvature of the surfaces on which the charged elements are situated and the area of the latter.

Judging from my past experience, as much as 100,000,000 volts are perfectly practicable. On the other hand currents of many thousands of amperes may be obtained in the antenna. A plant of but very moderate dimensions is required for such performances. Theoretically, a terminal of less than 90 feet in diameter is sufficient to develop an electromotive force of that magnitude while for antenna currents of from 2,000-4,000 amperes at the usual frequencies it need not be larger than 30 feet in diameter.

In a more restricted meaning this wireless transmitter is one in which the Hertz-wave radiation is an entirely negligible quantity as compared with the whole energy, under which condition the damping factor is extremely small and an enormous charge is stored in the elevated capacity. Such a circuit may then be excited with impulses of any kind, even of low frequency and it will yield sinusoidal and continuous oscillations like those of an alternator.

Taken in the narrowest significance of the term, however, it is a resonant transformer which, besides possessing these qualities, is accurately proportioned to fit the globe and its electrical constants and properties, by virtue of which design it becomes highly efficient and effective in the wireless transmission of energy. Distance is then absolutely eliminated, there being no diminution in the intensity of the transmitted impulses. It is even possible to make the actions increase with the distance from the plant according to an exact mathematical law.

This invention was one of a number comprised in my "World-System" of wireless transmission which I undertook to commercialize on my return to New York in 1900. As to the immediate purposes of my enterprise, they were clearly outlined in a technical statement of that period from which I quote:

"The 'World-System' has resulted from a combination of several original discoveries made by the inventor in the course of long continued research and experimentation. It makes possible not only the instantaneous and precise wireless transmission of any kind of signals, messages or characters, to all parts of the world, but also the inter-connection of the existing telegraph, telephone, and other signal stations without any change in their present equipment. By its means, for instance, a telephone subscriber here may call up and talk to any other subscriber on the Globe. An inexpensive receiver, not bigger than a watch, will enable him to listen anywhere, on land or sea, to a speech delivered or music played in some other place, however distant. These examples are cited merely to give an idea of the possibilities of this great scientific advance, which annihilates distance and makes that perfect natural conductor, the Earth, available for all the innumerable purposes which human ingenuity has found for a line-wire. One far-reaching result of this is that any device capable of being operated thru one or more wires (at a distance obviously restricted) can likewise be actuated, without artificial conductors and with the same facility and accuracy, at distances to which there are no limits other than those imposed by the physical dimensions of the Globe. Thus, not only will entirely new fields for commercial exploitation be opened up by this ideal method of transmission but the old ones vastly extended.

The 'World -System' is based on the application of the following important inventions and discoveries:

1. The 'Tesla Transformer.' This apparatus is in the production of electrical vibrations as revolutionary as gunpowder was in warfare. Currents many times stronger than any ever generated

in the usual ways, and sparks over one hundred feet long, have been produced by the inventor with an instrument of this kind.

2. The 'Magnifying Transmitter.' This is Tesla's best invention, a peculiar transformer specially adapted to excite the Earth, which is in the transmission of electrical energy what the telescope is in astronomical observation. By the use of this marvelous device he has already set up electrical movements of greater intensity than those of lightning and passed a current, sufficient to light more than two hundred incandescent lamps, around the Globe.

3. The 'Tesla Wireless System.' This system comprises a number of improvements and is the only means known for transmitting economically electrical energy to a distance without wires. Careful tests and measurements in connection with an experimental station of great activity, erected by the inventor in Colorado, have demonstrated that power in any desired amount can be conveyed, clear across the Globe if necessary, with a loss not exceeding a few per cent.

4. The 'Art of Individualization.' This invention of Tesla's is to primitive 'tuning' what refined language is to unarticulated expression. It makes possible the transmission of signals or messages absolutely secret and exclusive both in the active and passive aspect, that is, non- interfering as well as non-interferable. Each signal is like an individual of unmistakable identity and there is virtually no limit to the number of stations or instruments which can be simultaneously operated without the slightest mutual disturbance.

5. 'The Terrestrial Stationary Waves.' This wonderful discovery, popularly explained, means that the Earth is responsive to electrical vibrations of definite pitch just as a tuning fork to certain waves of sound. These particular electrical vibrations,

capable of powerfully exciting the Globe, lend themselves to innumerable uses of great importance commercially and in many other respects.

The first 'World -System' power plant can be put in operation in nine months. With this power plant it will be practicable to attain electrical activities up to ten million horsepower and it is designed to serve for as many technical achievements as are possible without due expense. Among these the following may be mentioned:

(1) The inter-connection of the existing telegraph exchanges or offices all over the world;

(2) The establishment of a secret and non-interferable government telegraph service;

(3) The inter-connection of all the present telephone exchanges or offices on the Globe;

(4) The universal distribution of general news, by telegraph or telephone, in connection with the Press;

(5) The establishment of such a 'World-System' of intelligence transmission for exclusive private use;

(6) The inter-connection and operation of all stock tickers of the world;

(7) The establishment of a 'World-System' of musical distribution, etc.;

(8) The universal registration of time by cheap clocks indicating the hour with astronomical precision and requiring no attention whatever;

(9) The world transmission of typed or handwritten characters, letters, checks, etc.;

(10) The establishment of a universal marine service enabling the navigators of all ships to steer perfectly without compass, to determine the exact location, hour and speed, to prevent collisions and disasters, etc.;

(11) The inauguration of a system of world-printing on land and sea;

(12) The world reproduction of photographic pictures and all kinds of drawings or records.

I also proposed to make demonstrations in the wireless transmission of power on a small scale but sufficient to carry conviction. Besides these I referred to other and incomparably more important applications of my discoveries which will be disclosed at some future date.

A plant was built on Long Island with a tower 187 feet high, having a spherical terminal about 68 feet in diameter. These dimensions were adequate for the transmission of virtually any amount of energy. Originally only from 200 to 300 K.W. were provided but I intended to employ later several thousand horsepower. The transmitter was to emit a wave complex of special characteristics and I had devised a unique method of telephonic control of any amount of energy.

The tower was destroyed two years ago but my projects are being developed and another one, improved in some features, will be constructed. On this occasion I would contradict the widely circulated report that the structure was demolished by the Government which owing to war conditions, might have created prejudice in the minds of those who may not know that the papers, which thirty years ago conferred upon me the honor of American citizenship, are always kept in a safe, while my orders, diplomas, degrees, gold medals and other distinctions are packed away in old trunks. If this report had a foundation I would have been refunded a large sum of money which I expended in the construction of the tower. On the contrary it was in the interest of the Government to preserve it, particularly as it would have made possible — to mention just one valuable result — the location of a submarine in any part of the world. My plant, services, and all my improvements have always been at the disposal of the officials and ever since the outbreak of the European conflict I have been working at a sacrifice on several inventions of mine relating to aerial navigation, ship propulsion and wireless transmission which are of the greatest importance to the country. Those who are well informed know that my ideas have revolutionized the industries of the United States and I am not aware that there lives an inventor who has been, in this respect, as fortunate as myself especially as regards the use of his improvements in the war. I have refrained from publicly expressing myself on this subject before as it seemed improper to dwell on personal matters while all the world was in dire trouble.

I would add further, in view of various rumors which have reached me, that Mr. J. Pierpont Morgan did not interest himself with me in a business way but in the same large spirit in which he has assisted many other pioneers. He carried out his generous promise to the letter and it would have been most unreasonable to expect from him anything more. He had the highest regard for my attainments and gave me every evidence of his complete faith in my ability to ultimately achieve what I had set out to do. I am unwilling to accord to some smallminded and jealous individuals the satisfaction of having thwarted my efforts. These men are to me nothing more than microbes of a nasty disease. My project was retarded by laws of nature. The world was not prepared for it. It was too far ahead of time. But the same laws will prevail in the end and make it a triumphal success.

VI. The Art of Telautomatics

No subject to which I have ever devoted myself has called for such concentration of mind and strained to so dangerous a degree the finest fibers of my brain as the system of which the Magnifying Transmitter is the foundation. I put all the intensity and vigor of youth in the development of the rotating field discoveries, but those early labors were of a different character. Although strenuous in the extreme, they did not involve that keen and exhausting discernment which had to be exercised in attacking the many puzzling problems of the wireless. Despite my rare physical endurance at that period the abused nerves finally rebelled and I suffered a complete collapse, just as the consummation of the long and difficult task was almost in sight.

Without doubt I would have paid a greater penalty later, and very likely my career would have been prematurely terminated, had not providence equipt me with a safety device, which has seemed to improve with advancing years and unfailingly comes into play when my forces are at an end. So long as it operates I am safe from danger, due to overwork, which threatens other inventors and, incidentally, I need no vacations which are indispensable to most people. When I am all but used up I simply do as the darkies, who "naturally fall asleep while white folks worry." To venture a theory out of my sphere, the body probably accumulates little by little a definite quantity of some toxic agent and I sink into a nearly lethargic state which lasts half an hour to the minute. Upon awakening I have the sensation as though the events immediately preceding had occurred very long ago, and if I attempt to continue the interrupted train of thought I feel a veritable mental nausea. Involuntarily I then turn to other work and am surprised at the freshness of the mind and ease with which I overcome obstacles that had baffled me before. After weeks or months my passion for the temporarily abandoned invention returns and I invariably find answers to all the vexing questions with scarcely any effort.

In this connection I will tell of an extraordinary experience which may be of interest to students of psychology. I had produced a striking phenomenon with my grounded transmitter and was endeavoring to ascertain its true significance in relation to the currents propagated through the earth. It seemed a hopeless undertaking, and for more than a year I worked unremittingly, but in vain. This profound study so entirely absorbed me that I became forgetful of everything else, even of my undermined health. At last, as I was at the point of breaking down, nature applied the preservative inducing lethal sleep. Regaining my senses I realized with consternation that I was unable to visualize scenes from my life except those of infancy, the very first ones that had entered my consciousness. Curiously enough, these appeared before my vision with startling distinctness and afforded me welcome relief. Night after night, when retiring, I would think of them and more and more of my previous existence was revealed. The image of my mother was always the principal figure in the spectacle that slowly unfolded, and a consuming desire to see her again gradually took possession of me. This feeling grew so strong that I resolved to drop all work and satisfy my longing. But I found it too hard to break away from the laboratory, and several months elapsed during which I had succeeded in reviving all the impressions of my past life up to the spring of 1892. In the next picture that came out of the mist of oblivion, I saw myself at the Hotel de la Paix in Paris just coming to from one of my peculiar sleeping spells, which had been caused by prolonged exertion of the brain. Imagine the pain and distress I felt when it flashed upon my mind that a dispatch was handed to me at that very moment bearing the sad news that my mother was dying. I remembered how I made the long journey home without an hour of rest and how she passed away after weeks of agony! It was especially remarkable that during all this period of partially obliterated memory I was fully alive to everything touching on the subject of my research. I could recall the smallest details and the least significant observations in my experiments and even recite pages of text and complex mathematical formulae.

My belief is firm in a law of compensation. The true rewards are ever in proportion to the labor and sacrifices made. This is one of the reasons why I feel certain that of all my inventions, the Magnifying Transmitter will prove most important and valuable to future generations. I am prompted to this prediction not so much by thoughts of the commercial and industrial revolution which it will surely bring about, but of the humanitarian consequences of the many achievements it makes possible. Considerations of mere utility weigh little in the balance against the higher benefits of civilization. We are confronted with portentous problems which can not be solved just by providing for our material existence, however abundantly. On the contrary, progress in this direction is fraught with hazards and perils not less menacing than those born from want and suffering. If we were to release the energy of atoms or discover some other way of developing cheap and unlimited power at any point of the globe this accomplishment, instead of being a blessing, might bring disaster to mankind in giving rise to dissension and anarchy which would ultimately result in the enthronement of the hated regime of force. The greatest good will comes from technical improvements tending to unification and harmony, and my wireless transmitter is preeminently such. By its means the human voice and likeness will be reproduced everywhere and factories driven thousands of miles from waterfalls furnishing the power; aerial machines will be propelled around the earth without a stop and the sun's energy controlled to create lakes and rivers for motive purposes and transformation of arid deserts into fertile land. Its introduction for telegraphic, telephonic and similar uses will automatically cut out the statics and all other interferences which at present impose narrow limits to the application of the wireless.

This is a timely topic on which a few words might not be amiss. During the past decade a number of people have arrogantly claimed that they had succeeded in doing away with this impediment. I have carefully examined all of the arrangements described and tested most of them long before they were publicly disclosed, but the finding was uniformly negative. A recent official statement from the U.S. Navy may, perhaps, have taught some beguilable news editors how to appraise these announcments at their real worth. As a rule the attempts are based on theories so fallacious that whenever they come to my notice I can not help thinking in a lighter vein. Quite recently a new discovery was heralded, with a deafening flourish of trumpets, but it proved another case of a mountain bringing forth a mouse.

This reminds me of an exciting incident which took place years ago when I was conducting my experiments with currents of high frequency. Steve Brodie had just jumped off the Brooklyn Bridge. The feat has been vulgarized since by imitators, but the first report electrified New York. I was very impressionable then and frequently spoke of the daring printer. On a hot afternoon I felt the necessity of refreshing myself and stepped into one of the popular thirty thousand institutions of this great city where a delicious twelve per cent beverage was served which can now be had only by making a trip to the poor and devastated countries of Europe. The attendance was large and not overdistinguished and a matter was discussed which gave me

67

an admirable opening for the careless remark: "This is what I said when I jumped off the bridge." No sooner had I uttered these words than I felt like the companion of Timotheus in the poem of Schiller. In an instant there was a pandemonium and a dozen voices cried: "It is Brodie! " I threw a quarter on the counter and bolted for the door but the crowd was at my heels with yells: "Stop, Steve!" which must have been misunderstood for many persons tried to hold me up as I ran frantically for my haven of refuge. By darting around corners I fortunately managed - through the medium of a fire-escape - to reach the laboratory where I threw off my coat, camouflaged myself as a hard-working blacksmith, and started the forge. But these precautions proved unnecessary; I had eluded my pursuers. For many years afterward, at night, when imagination turns into spectres the trifling troubles of the day, I often thought, as I tossed on the bed, what my fate would have been had that mob caught me and found out that I was not Steve Brodie!

Now the engineer, who lately gave an account before a technical body of a novel remedy against statics based on a "heretofore unknown law of nature," seems to have been as reckless as myself when he contended that these disturbances propagate up and down, while those of a transmitter proceed along the earth. It would mean that a condenser, as this globe, with its gaseous envelope, could be charged and discharged in a manner quite contrary to the fundamental teachings propounded in every elemental text-book of physics. Such a supposition would have been condemned as erroneous, even in Franklin's time, for the facts bearing on this were then well known and the identity between atmospheric electricity and that developed by machines was fully established. Obviously, natural and artificial disturbances propagate through the earth and the air in exactly the same way, and both set up electromotive forces in the horizontal, as well as vertical, sense. Interference can not be overcome by any such methods as were proposed. The truth is this: in the air the potential increases at the rate of about fifty volts per foot of elevation, owing to which there may be a difference of pressure amounting to twenty, or even forty thousand volts between the upper and lower ends of the antenna. The masses of the charged atmosphere are constantly in motion and give up electricity to the conductor, not continuously but rather disruptively, this producing a grinding noise in a sensitive telephonic receiver. The higher the terminal and the greater the space encompassed by the wires, the more pronounced is the effect, but it must be understood that it is purely local and has little to do with the real trouble.

In 1900, while perfecting my wireless system, one form of apparatus comprised four antennae. These were carefully calibrated to the same frequency and connected in multiple with the object of magnifying the action, in receiving from any direction. When I desired to ascertain the origin of the transmitted impulses, each diagonally situated pair was put in series with a primary coil energizing the detector circuit. In the former case the sound was loud in the telephone; in the latter it ceased, as expected, the two antennae neutralizing each other, but the true statics manifested themselves in both instances and I had to devise special preventives embodying different principles.

By employing receivers connected to two points of the ground, as suggested by me long ago, this trouble caused by the charged air, which is very serious in the structures as now built, is nullified and besides, the liability of all kinds of interference is reduced to about one-half, because of the directional character of the circuit. This was perfectly self-evident, but came as a revelation to some simple-minded wireless folks whose experience was confined to forms of apparatus that could have been improved with an axe, and they have been disposing of the bear's skin before killing him. If it were true that strays performed such antics, it would be easy to get rid of them by receiving without aerials. But, as a matter of fact, a wire buried in the ground which, conforming to this view, should be absolutely immune, is more susceptible to certain extraneous impulses than one placed vertically in the air. To state it fairly, a slight progress has been made, but not by virtue of any particular method or device. It was achieved simply by discarding the enormous structures, which are bad enough for transmission but wholly unsuitable for reception, and adopting a more appropriate type of receiver. As I pointed out in a previous article, to dispose of this difficulty for good, a radical change must be made in the system, and the sooner this is done the better.

It would be calamitous, indeed, if at this time when the art is in its infancy and the vast majority, not excepting even experts, have no conception of its ultimate possibilities, a measure would be rushed through the legislature making it a government monopoly. This was proposed a few weeks ago by Secretary Daniels, and no doubt that distinguished official has made his appeal to the Senate and House of Representatives with sincere conviction. But universal evidence unmistakably shows that the best results are always obtained in healthful commercial competition. There are, however, exceptional reasons why wireless should be given the fullest freedom of development. In the first place it offers prospects immeasurably greater and more vital to betterment of human life than any other invention or discovery in the history of man. Then again, it must be understood that this wonderful art has been, in its entirety, evolved here and can be called "American" with more right and propriety than the telephone, the incandescent lamp or the aeroplane. Enterprising press agents and stock jobbers have been so successful in spreading misinformation that even so excellent a periodical as the Scientific American accords the chief credit to a foreign country. The Germans, of course, gave us the Hertz-waves and the Russian, English, French and Italian experts were quick in using them for signaling purposes. It was an obvious application of the new agent and accomplished with the old classical and unimproved induction coil - scarcely anything more than another kind of heliography. The radius of transmission was very limited, the results attained of little value, and the Hertz oscillations, as a means for conveying intelligence, could have been advantageously replaced by sound-waves, which I advocated in 1891. Moreover, all of these attempts were made three years after the basic principles of the wireless system, which is universally employed to-day, and its potent instrumentalities had been clearly described and developed in America. No trace of those Hertzian appliances and methods remains today. We have proceeded in the very opposite direction and what has been done is the product of the brains and efforts of citizens of this country. The fundamental patents have expired and the opportunities are open to all. The chief argument of the Secretary is based on interference. According to his statement, reported in the New York Herald of July 29th, signals from a powerful station can be intercepted in every village of the world . In view of this fact, which was demonstrated in my experiments of 1900, it would be of little use to impose restrictions in the United States.

As throwing light on this point, I may mention that only recently an odd looking gentleman called on me with the object of enlisting my services in the construction of world transmitters in some distant land. "We have no money," he said, "but carloads of solid gold and we will give you a liberal amount." I told him that I wanted to see first what will be done with my inventions in America, and this ended the interview. But I am satisfied that some dark forces are at work, and as time goes on the maintenance of continuous communication will be rendered more difficult. The only remedy is a system immune against interruption. It has been perfected, it exists, and all that is necessary is to put it in operation.

The terrible conflict is still uppermost in the minds and perhaps the greatest importance will be attached to the Magnifying Transmitter as a machine for attack and defense, more particularly in connection with Telautomatics. This invention is a logical outcome of observations begun in my boyhood and continued thruout my life. When the first results were publisht the Electrical Review stated editorially that it would become one of the "most potent factors in the advance and civilization of mankind." The time is not distant when this prediction will be fulfilled. In 1898 and 1900 it was offered to the Government and might have been adopted were I one of those who would go to Alexander's shepherd when they want a favor from Alexander. At that time I really thought that it would abolish war, because of its unlimited destructiveness and exclusion of the personal element of combat. But while I have not lost faith in its potentialities, my views have changed since.

War can not be avoided until the physical cause for its recurrence is removed and this, in the last analysis, is the vast extent of the planet on which we live. Only thru annihilation of distance in every respect, as the conveyance of intelligence, transport of passengers and supplies and transmission of energy will conditions be brought about some day, insuring permanency of friendly relations. What we now want most is closer contact and better understanding between individuals and communities all over the earth, and the elimination of that fanatic devotion to exalted ideals of national egoism and pride which is always prone to plunge the world into primeval barbarism and strife. No league or parliamentary act of any kind will ever prevent such a calamity. These are only new devices for putting the weak at the mercy of the strong. I have exprest myself in this regard fourteen years ago, when a combination of a few leading governments - a sort of Holy Alliance - was advocated by the late Andrew Carnegie, who may be fairly considered as the father of this idea, having given to it more publicity and impetus than anybody else prior to the efforts of the President. While it can not be denied that such a pact might be of material advantage to some less fortunate peoples, it can not attain the chief object sought. Peace can only come as a natural consequence of universal enlightenment and merging of races, and we are still far from this blissful realization.

As I view the world of today, in the light of the gigantic struggle we have witnest, I am filled with conviction that the interests of humanity would be best served if the United States remained true to its traditions and kept out of "entangling alliances." Situated as it is, geographically, remote from the theaters of impending conflicts, without incentive to territorial aggrandizement, with inexhaustible resources and immense population thoroly imbued with the spirit of liberty and right, this country is placed in a unique and privileged position. It is thus able to exert, independently, its colossal strength and moral force to the benefit of all, more judiciously and effectively, than as member of a league.

In one of these biographical sketches, published in the ELECTRICAL EXPERIMENTER, I have dwelt on the circumstances of my early life and told of an affliction which compelled me to unremitting exercise of imagination and self observation. This mental activity, at first involuntary under the pressure of illness and suffering, gradually became second nature and led me finally to recognize that I was but an automaton devoid of free will in thought and action and merely responsive to the forces of the environment. Our bodies are of such complexity of structure, the motions we perform are so numerous and involved, and the external impressions on our sense organs to such a degree delicate and elusive that it is hard for the average person to grasp this fact. And yet nothing is more convincing to the trained investigator than the mechanistic theory of life which had been, in a measure, understood and propounded by Descartes three hundred years ago. But in his time many important functions of our organism were unknown and, especially with respect to the nature of light and the construction and operation of the eye, philosophers were in the dark.

In recent years the progress of scientific research in these fields has been such as to leave no room for a doubt in regard to this view on which many works have been published. One of its ablest and most eloquent exponents is, perhaps, Felix Le Dantec, formerly assistant of Pasteur. Prof. Jacques Loeb has performed remarkable experiments in heliotropism, clearly establishing the controlling power of light in lower forms of organisms, and his latest book, "Forced Movements," is revelatory. But while men of science accept this theory simply as any other that is recognized, to me it is a truth which I hourly demonstrate by every act and thought of mine. The consciousness of the external impression prompting me to any kind of exertion, physical or mental, is ever present in my mind. Only on very rare occasions, when I was in a state of exceptional concentration, have I found difficulty in locating the original impulses.

The by far greater number of human beings are never aware of what is passing around and within them, and millions fall victims of disease and die prematurely just on this account. The commonest every -day occurrences appear to them mysterious and inexplicable. One may feel a sudden wave of sadness and rake his brain for an explanation when he might have noticed that it was caused by a cloud cutting off the rays of the sun. He may see the image of a friend dear to him under conditions which he construes as very peculiar, when only shortly before he has passed him in the street or seen his photograph somewhere. When he loses a collar button he fusses and swears for an hour, being unable to visualize his previous actions and locate the object directly. Deficient observation is merely a form of ignorance and responsible for the many morbid notions and foolish ideas prevailing. There is not more than one out of every ten persons who does not believe in telepathy and other psychic manifestations, spiritualism and communion with the dead, and who would refuse to listen to willing or unwilling deceivers.

Just to illustrate how deeply rooted this tendency has become even among the clearheaded American population, I may mention a comical incident. Shortly before the war, when the exhibition of my turbines in this city elicited widespread comment in the technical papers, I anticipated that there would, be a scramble among manufacturers to get hold of the invention, and I had particular designs on that man from Detroit who has an uncanny faculty for accumulating millions. So confident was I that he would turn up some day, that I declared this as certain to my secretary and assistants. Sure enough, one fine morning a body of engineers from the Ford Motor Company presented themselves with the request of discussing with me an important project. "Didn't I tell you?" I remarked triumphantly to my employees, and one of them said, "You are amazing, Mr. Tesla; everything comes out exactly as you predict." As soon as these hard-headed men were seated I, of course, immediately began to extol the wonderful features of my turbine, when the spokesmen interrupted me and said, "We know all about this, but we are on a special errand. We have formed a psychological society for the investigation of psychic phenomena and we want you to join us in this undertaking." I suppose those engineers never knew how near they came to being fired out of my office.

Ever since I was told by some of the greatest men of the time, leaders in science whose names are immortal, that I am possesst of an unusual mind, I bent all my thinking faculties on the solution of great problems regardless of sacrifice. For many years I endeavored to solve the enigma of death, and watched eagerly for every kind of spiritual indication. But only once in the course of my existence have I had an experience which momentarily impressed me as supernatural. It was at the time of my mother's death. I had become completely exhausted by pain and long vigilance, and one night was carried to a building about two blocks from our home. As I lay helpless there, I thought that if my mother died while I was away from her bedside she would surely give me a sign. Two or three months before I was in London in company with my late friend, Sir William Crookes, when spiritualism was discussed, and I was under the full sway of these thoughts. I might not have paid attention to other men, but was susceptible to his arguments as it was his epochal work on radiant matter, which I had read as a student, that made me embrace the electrical career. I reflected that the conditions for a look into the beyond were most favorable, for my mother was a woman of genius and particularly excelling in the powers of intuition. During the whole night every fiber in my brain was strained in expectancy, but nothing happened until early in the morning, when I fell in a sleep, or perhaps a swoon, and saw a cloud carrying angelic figures of marvelous beauty, one of whom gazed upon me lovingly and gradually assumed the features of my mother. The appearance slowly floated across the room and vanished, and I was awakened by an indescribably sweet song of many voices. In that instant a certitude, which no words can express, came upon me that my mother had just died.

76

And that was true. I was unable to understand the tremendous weight of the painful knowledge I received in advance, and wrote a letter to Sir William Crookes while still under the domination of these impressions and in poor bodily health. When I recovered I sought for a long time the external cause of this strange manifestation and, to my great relief, I succeeded after many months of fruitless effort. I had seen the painting of a celebrated artist, representing allegorically one of the seasons in the form of a cloud with a group of angels which seemed to actually float in the air, and this had struck me forcefully. It was exactly the same that appeared in my dream, with the exception of my mother's likeness. The music came from the choir in the church nearby at the early mass of Easter morning, explaining everything satisfactorily in conformity with scientific facts.

This occurred long ago, and I have never had the faintest reason since to change my views on psychical and spiritual phenomena, for which there is absolutely no foundation. The belief in these is the natural outgrowth of intellectual development. Religious dogmas are no longer accepted in their orthodox meaning, but every individual clings to faith in a supreme power of some kind. We all must have an ideal to govern our conduct and insure contentment, but it is immaterial whether it be one of creed, art, science or anything else, so long as it fulfills the function of a dematerializing force. It is essential to the peaceful existence of humanity as a whole that one common conception should prevail.

While I have failed to obtain any evidence in support of the contentions of psychologists and spiritualists, I have proved to my complete satisfaction the automatism of life, not only through continuous observations of individual actions, but even more conclusively through certain generalizations. These amount to a discovery which I consider of the greatest moment to human society, and on which I shall briefly dwell. I got the first inkling of this astounding truth when I was still a very young man, but for many years I interpreted what I noted simply as coincidences. Namely, whenever either myself or a person to whom I was attached, or a cause to which I was devoted, was hurt by others in a particular way, which might be best popularly characterized as the most unfair imaginable, I experienced a singular and undefinable pain which, for want of a better term, I have qualified as "cosmic," and shortly thereafter, and invariably, those who had inflicted it came to grief. After many such cases I confided this to a number of friends, who had the opportunity to convince themselves of the truth of the theory which I have gradually formulated and which may be stated in the following few words:

Our bodies are of similar construction and exposed to the same external influences. This results in likeness of response and concordance of the general activities on which all our social and other rules and laws are based. We are automata entirely controlled by the forces of the medium being tossed about like corks on the surface of the water, but mistaking the resultant of the impulses from the outside for free will. The movements and other actions we perform are always life preservative and tho seemingly quite independent from one another, we are connected by invisible links. So long as the organism is in perfect order it responds accurately to the agents that prompt it, but the moment that there is some derangement in any individual, his self- preservative power is impaired. Everybody understands, of course, that if one becomes deaf, has his eyesight weakened, or his limbs injured, the chances for his continued existence are lessened. But this is also true, and perhaps more so, of certain defects in the brain which deprive the automaton, more or less, of that vital quality and cause it to rush into destruction. A very sensitive and observant being, with his highly developed mechanism all intact, and acting with precision in obedience to the changing conditions of the environment, is endowed with a transcending mechanical sense, enabling him to evade perils too subtle to be directly perceived. When he comes in contact with others whose controlling organs are radically faulty, that sense asserts itself and he feels the "cosmic" pain. The truth of this has been borne out in hundreds of instances and I am inviting other students of nature to devote attention to this subject, believing that thru combined and systematic effort results of incalculable value to the world will be attained.

The idea of constructing an automaton, to bear out my theory, presented itself to me early but I did not begin active work until 1893, when I started my wireless investigations. During the succeeding two or three years a number of automatic mechanisms, to be actuated from a distance, were constructed by me and exhibited to visitors in my laboratory. In 1896, however, I designed a complete machine capable of a multitude of operations, but the consummation of my labors was delayed until late in 1897. This machine was illustrated and described in my article in the Century Magazine of June, 1900, and other periodicals of that time and, when first shown in the beginning of 1898, it created a sensation such as no other invention of mine has ever produced. In November, 1898, a basic patent on the novel art was granted to me, but only after the Examiner-in-Chief had come to New York and witnesst the performance, for what I claimed seemed unbelievable. I remember that when later I called on an official in Washington, with a view of offering the invention to the Government, he burst out in laughter upon my telling him what I had accomplished. Nobody thought then that there was the faintest prospect of perfecting such a device. It is unfortunate that in this patent, following the advice of my attorneys, I indicated the control as being effected thru the medium of a single circuit and a well-known form of detector, for the reason that I had not yet secured protection on my methods and apparatus for individualization. As a matter of fact, my boats were controlled thru the joint action of several circuits and interference of every kind was excluded. Most generally I employed receiving circuits in the form of loops, including condensers, because the discharges of my high-tension transmitter ionized the air in the hall so that even a very small aerial would draw electricity from the surrounding atmosphere for hours. Just to give an idea, I found, for instance, that a bulb 12" in diameter, highly exhausted, and with one single terminal to which a short wire was attached, would deliver well on to one thousand successive flashes before all charge of the air in the laboratory was neutralized. The loop form of receiver was not sensitive to such a disturbance and it is curious to note that it is becoming popular at this late date. In reality it collects much less energy than the aerials or a long grounded wire, but it so happens that it does away with a number of defects inherent to the present wireless devices. In demonstrating my invention before audiences, the visitors were requested to ask any questions, however involved, and the automaton would answer them by signs. This was considered magic at that time but was extremely simple, for it was myself who gave the replies by means of the device.

At the same period another larger telautomatic boat was constructed a photograph of which is shown in this number of the ELECTRICAL EXPERIMENTER. It was controlled by loops, having several turns placed in the hull, which was made entirely water-tight and capable of submergence. The apparatus was similar to that used in the first with the exception of certain special features I introduced as, for example, incandescent lamps which afforded a visible evidence of the proper functioning of the machine.

These automata, controlled within the range of vision of the operator, were, however, the first and rather crude steps in the evolution of the Art of Telautomatics as I had conceived it. The next logical improvement was its application to automatic mechanisms beyond the limits of vision and at great distance from the center of control, and I have ever since advocated their employment as instruments of warfare in preference to guns. The importance of this now seems to be recognized, if I am to judge from casual announcements thru the press of achievements which are said to be extraordinary but contain no merit of novelty, whatever. In an imperfect manner it is practicable, with the existing wireless plants, to launch an aeroplane, have it follow a certain approximate course, and perform some operation at a distance of many hundreds of miles. A machine of this kind can also be mechanically controlled in several ways and I have no doubt that it may prove of some usefulness in war. But there are, to my best knowledge, no instrumentalities in existence today with which such an object could be accomplished in a precise manner. I have devoted years of study to this matter and have evolved means, making such and greater wonders easily realizable.

As stated on a previous occasion, when I was a student at college I conceived a flying machine quite unlike the present ones. The underlying principle was sound but could not be carried into practice for want of a prime-mover of sufficiently great activity. In recent years I have successfully solved this problem and am now planning aerial machines devoid of sustaining planes, ailerons, propellers and other external attachments, which will be capable of immense speeds and are very likely to furnish powerful arguments for peace in the near future. Such a machine, sustained and propelled entirely by reaction, is shown on page 108 and is supposed to be controlled either mechanically or by wireless energy. By installing proper plants it will be

81

practicable to project a missile of this kind into the air and drop it almost on the very spot designated, which may be thousands of miles away. But we are not going to stop at this. Telautomata will be ultimately produced, capable of acting as if possest of their own intelligence, and their advent will create a revolution. As early as 1898 I proposed to representatives of a large manufacturing concern the construction and public exhibition of an automobile carriage which, left to itself, would perform a great variety of operations involving something akin to judgment. But my proposal was deemed chimerical at that time and nothing came from it.

At present many of the ablest minds are trying to devise expedients for preventing a repetition of the awful conflict which is only theoretically ended and the duration and main issues of which I have correctly predicted in an article printed in the Sun of December 20, 1914. The proposed League is not a remedy but on the contrary, in the opinion of a number of competent men, may bring about results just the opposite. It is particularly regrettable that a punitive policy was adopted in framing the terms of peace, because a few years hence it will be possible for nations to fight without armies, ships or guns, by weapons far more terrible, to the destructive action and range of which there is virtually no limit. A city, at any distance whatsoever from the enemy, can be destroyed by him and no power on earth can stop him from doing so. If we want to avert an impending calamity and a state of things which may transform this globe into an inferno, we should push the development of flying machines and wireless transmission of energy without an instant's delay and with all the power and resources of the nation.

THE TRUE WIRELESS

by <u>Nikola Tesla</u>

<u>Electrical Experimenter, The</u>

May, 1919

Page number(s):

28-30, 61-63 & 87

In this remarkable and complete story of his discovery of the "True Wireless" and the principles upon which transmission and reception, even in the present day systems, are based, Dr. Nikola Tesla shows us that he is indeed the "Father of the Wireless." To him the Hertz wave theory is a delusion; it looks sound from certain angles, but the facts tend to prove that it is hollow and empty. He convinces us that the real Hertz waves are blotted out after they have traveled but a short distance from the sender. It follows, therefore, that the measured antenna current is no indication of the effect, because only a small part of it is effective at a distance. The limited activity of pure Hertz wave transmission and reception is here clearly explained, besides showing definitely that in spite of themselves, the radio engineers of today are employing the original Tesla tuned oscillatory system. He shows by examples with different forms of aërials that the signals picked up by the instruments must actually be induced by earth currents — not etheric space waves. Tesla also disproves the "Heaviside layer" theory from his personal observations and tests.
EDITOR.

Fig. 1. — Alternator of 10,000 Cycles p.s., Capacity 10 K.W., Which Was Employed by Tesla in His First Demonstrations of High Frequency Phenomena Before the American Institute of Electrical Engineers at Columbia College, May 20, 1891.

Ever since the announcement of Maxwell's electro-magnetic theory scientific investigators all the world over had been bent on its experimental verification. They were convinced that it would be done and lived in an atmosphere of eager expectancy, unusually favorable to the reception of any evidence to this end. No wonder then that the publication of Dr. Heinrich Hertz's results caused a thrill as had scarcely ever been experienced before. At that time I was in the midst of pressing work in connection with the commercial introduction of my system of power transmission, but, nevertheless, caught the fire of enthusiasm and fairly burned with desire to behold the miracle with my own eyes. Accordingly, as soon as I had freed myself of these imperative duties and resumed research work in my laboratory on Grand Street, New York, I began, parallel with high frequency alternators, the construction of several forms of apparatus with the object of exploring the field opened up by Dr. Hertz. Recognizing the limitations of the devices he had employed, I concentrated my attention on the production of a powerful induction coil but made no notable progress until a happy inspiration led me to the invention of the oscillation transformer. In the latter part of 1891 I was already so far advanced in the development of this new principle that I had at my disposal means vastly superior to those of the German physicist. All my previous efforts with Rhumkorf coils had left me unconvinced, and in order to settle my doubts I went over the whole ground once more, very carefully, with these improved appliances. Similar phenomena were noted, greatly magnified in intensity, but they were susceptible of a different and more plausible explanation. I considered this so important that in 1892 I went to Bonn, Germany, to confer with Dr. Hertz in regard to my observations. He seemed disappointed to such a degree that I regretted my trip and parted from him sorrowfully. During the succeeding years I made numerous experiments with the same object, but the results were uniformly negative. In 1900, however, after I had evolved a wireless transmitter which enabled me to obtain electro-magnetic activities of many millions of horse-power, I made a last desperate attempt to prove that the disturbances emanating from the oscillator were ether vibrations akin to those of light, but met again with utter failure. For more than eighteen years I have been reading treatises, reports of scientific transactions, and articles on Hertz-wave telegraphy, to keep myself informed, but they have always imprest me like works of fiction.

Fig. 2. — Diagram Illustrating the Circuit Connections and Tuning Devices Employed by Tesla In His Experimental Demonstrations Before the American Institute of Electrical Engineers With the High Frequency Alternator Shown in Fig. 1.

The history of science shows that theories are perishable. With every new truth that is revealed we get a better understanding of Nature and our conceptions and views are modified. Dr. Hertz did not discover a new principle. He merely gave material support to a hypothesis which had been long ago formulated. It was a perfectly well-established fact that a circuit, traversed by a periodic current, emitted some kind of space waves, but we were in ignorance as to their character. He apparently gave an experimental proof that they were transversal vibrations in the ether. Most people look upon this as his great accomplishment. To my mind it seems that his immortal merit was not so much in this as in the focusing of the investigators' attention on the processes taking place in the ambient medium. The Hertz-wave theory, by its fascinating hold on the imagination, has stifled creative effort in the wireless art and retarded it for twenty-five years. But, on the other hand, it is impossible to over-estimate the beneficial effects of the powerful stimulus it has given in many directions.

Fig. 3. — Electric Transmission Thru Two Wires and Hydraulic Analog.

As regards signaling without wires, the application of these radiations for the purpose was quite obvious. When Dr. Hertz was asked whether such a system would be of practical value, he did not think so, and he was correct in his forecast. The best that might have been expected was a method of communication similar to the heliographic and subject to the same or even greater limitations.

In the spring of 1891 I gave my demonstrations with a high frequency machine before the American Institute of Electrical Engineers at Columbia College, which laid the foundation to a new and far more promising departure. Altho the laws of electrical resonance were well known at that time and my lamented friend, Dr. John Hopkinson, had even indicated their specific application to an alternator in the Proceedings of the Institute of Electrical Engineers, London, Nov. 13, 1889, nothing had been done towards the practical use of this knowledge and it is probable that those experiments of mine were the first public exhibition with resonant circuits, more particularly of high frequency. While the spontaneous success of my lecture was due to spectacular features, its chief import was in showing that all kinds of devices could be operated thru a single wire without return. This was the initial step in the evolution of my wireless system. The idea presented itself to me that it might be possible, under observance of proper conditions of resonance, to transmit electric energy thru the earth, thus dispensing with all artificial conductors. Anyone who might wish to examine impartially the merit of that early suggestion must not view it in the light of present day science. I only need to say that as late as 1893, when I had prepared an elaborate chapter on my wireless system, dwelling on its various instrumentalities and future prospects, Mr. Joseph Wetzler and other friends of mine emphatically protested against its publication on the ground that such idle and far-fetched speculations would injure me in the opinion of conservative business men. So it came that only a small part of what I had intended to say was embodied in my address of that year before the Franklin Institute and National Electric Light Association under the chapter "On Electrical Resonance." This little salvage from the wreck has earned me the title of "Father of the Wireless" from many well-disposed fellow workers, rather than the invention of scores of appliances which have brought wireless transmission within the reach of every young amateur and which, in a time not distant, will lead to undertakings overshadowing in magnitude and importance all past achievements of the engineer.

Fig. 4. — Electric Transmission Thru a Single Wire Hydraulic Analog.

The popular impression is that my wireless work was begun in 1893, but as a matter of fact I spent the two preceding years in investigations, employing forms of apparatus, some of which were almost like those of today. It was clear to me from the very start that the successful consummation could only be brought about by a number of radical improvements. Suitable high frequency generators and electrical oscillators had first to be produced. The energy of these had to be transformed in effective transmitters and collected at a distance in proper receivers. Such a system would be manifestly circumscribed in its usefulness if all extraneous interference were not prevented and exclusiveness secured. In time, however, I recognized that devices of this kind, to be most effective and efficient, should be designed with due regard to the physical properties of this planet and the electrical conditions obtaining on the same. I will briefly touch upon the salient advances as they were made in the gradual development of the system.

Fig. 5. — Illustrating Typical Arrangements for Collecting Energy In a System of Transmission Thru a Single Wire.

The high frequency alternator employed in my first demonstrations is illustrated in Fig. 1. It comprised a field ring, with 384 pole projections and a disc armature with coils wound in one single layer which were connected in various ways according to requirements. It was an excellent machine for experimental purposes, furnishing sinusoidal currents of from 10,000 to 20,000 cycles per second. The output was comparatively large, due to the fact that as much as 30 amperes per square millimeter could be past thru the coils without injury.

The diagram in Fig. 2 shows the circuit arrangements as used in my lecture. Resonant conditions were maintained by means of a condenser subdivided into small sections, the finer adjustments being effected by a movable iron core within an inductance coil. Loosely linked with the latter was a high tension secondary which was tuned to the primary.

The operation of devices thru a single wire without return was puzzling at first because of its novelty, but can be readily explained by suitable analogs. For this purpose reference is made to Figs. 3 and 4.

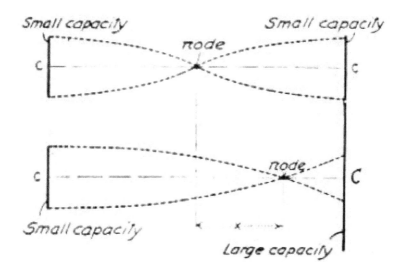

Fig. 6. — Diagram Elucidating Effect of Large Capacity on One End.

In the former the low resistance electrical conductors are represented by pipes of large section, the alternator by an oscillating piston and the filament of an incandescent lamp by a minute channel connecting the pipes. It will be clear from a glance at the diagram that very slight excursions of the piston would cause the fluid to rush with high velocity thru the small channel and that virtually all the energy of movement would be transformed into heat by friction, similarly to that of the electric current in the lamp filament.

The second diagram will now be self-explanatory. Corresponding to the terminal capacity of the electric system an elastic reservoir is employed which dispenses with the necessity of a return pipe. As the piston oscillates the bag expands and contracts, and the fluid is made to surge thru the restricted passage with great speed, this resulting in the generation of heat as in the incandescent lamp. Theoretically considered, the efficiency of conversion of energy should be the same in both cases.

Fig. 7. — Transmission of Electrical Energy Thru the Earth as Illustrated in Tesla's Lectures Before the Franklin Institute and Electric Light Association in February and March, 1893, and Mechanical Analog of the Same.

Granted, then, that an economic system of power transmission thru a single wire is practicable, the question arises how to collect the energy in the receivers. With this object attention is called to Fig. 5, in which a conductor is shown excited by an oscillator joined to it at one end. Evidently, as

94

the periodic impulses pass thru the wire, differences of potential will be created along the same as well as at right angles to it in the surrounding medium and either of these may be usefully applied. Thus at *a*, a circuit comprising an inductance and capacity is resonantly excited in the transverse, and at *b*, in the longitudinal sense. At *c*, energy is collected in a circuit parallel to the conductor but not in contact with it, and again at *d*, in a circuit which is partly sunk into the conductor and may be, or not, electrically connected to the same. It is important to keep these typical dispositions in mind, for however the distant actions of the oscillator might be modified thru the immense extent of the globe the principles involved are the same.

Fig. 8. — Tesla's System of Wireless Transmission Thru the Earth as Actually Exposed In His Lectures Before the Franklin Institute and Electric Light Association in February and March, 1893.

Consider now the effect of such a conductor of vast dimensions on a circuit exciting it. The upper diagram of Fig. 6 illustrates a familiar oscillating system comprising a straight rod of self-inductance $2L$ with small terminal capacities cc and a node in the center. In the lower diagram of the figure a large capacity C is attached to the rod at one end with the result of shifting the node to the right, thru a distance corresponding to self-inductance X. As both parts of the system on either side of the node vibrate at the same rate, we have evidently, $(L + X) c = (L - X) C$ from which $X = L (C - c / C + c)$. When the capacity C becomes commensurate to that of the earth, X approximates L, in other words, the node is close to the ground connection. *The exact determination of its position is very important in the calculation of certain terrestrial electrical and geodetic data* and I have devised special means with this purpose in view.

Fig. 9. — The Forerunner of the Audion — the Most Sensitive Wireless Detector Known, as Described by Tesla In His Lecture Before the Institution of Electrical Engineers, London, February, 1892.

My original plan of transmitting energy without wires is shown in the upper diagram of Fig. 7, while the lower one illustrates its mechanical analog, first publisht in my article in the *Century Magazine* of June, 1900. An alternator, preferably of high tension, has one of its terminals connected to the ground and the other to an elevated capacity and impresses its oscillations upon the earth. At a distant point a receiving circuit, likewise connected to ground and to an elevated capacity, collects some of the energy and actuates a suitable device. I suggested a multiplication of such units in order to intensify the effects, an idea which may yet prove valuable. In the analog two tuning forks are provided, one at the sending and the other at the receiving station, each having attached to its lower prong a piston fitting in a cylinder. The two cylinders communicate with a large elastic reservoir filled with an incompressible fluid. The vibrations transmitted to either of the tuning forks excite them by resonance and, thru electrical contacts or otherwise, bring about the desired result. This, I may say, was not a mere mechanical illustration, but a simple representation of my apparatus for submarine signaling, perfected by me in 1892, but not appreciated at that time, altho more efficient than the instruments now in use.

Fig. 10. — Tesla's System of Concatenated Tuned Circuits Shown and Described In U. S. Patent No. 568,178 of September 22, 1896, and Corresponding Arrangements in Wireless Transmission.

The electric diagram in Fig. 7, which was reproduced from my lecture, was meant only for the exposition of the principle. The arrangement, as I described it in detail, is shown in Fig. 8. In this case an alternator energizes the primary of a transformer, the high tension secondary of which is connected to the ground and an elevated capacity and tuned to the imprest oscillations. The receiving circuit consists of an inductance connected to the ground and to an elevated terminal without break and is resonantly responsive to the transmitted oscillations. A specific form of receiving device was not mentioned, but I had in mind to transform the received currents and thus make their volume and tension suitable for any purpose. This, in substance, is the system of today and I am not aware of a single authenticated instance of successful transmission at considerable distance by different instrumentalities. It might, perhaps, not be clear to those who have perused my first description of these improvements that, besides making known new and efficient types of apparatus, I gave to the world a wireless system of potentialities far beyond anything before conceived. I made explicit and repeated statements that I contemplated transmission, absolutely unlimited as to terrestrial distance and amount of energy. But, altho I have overcome all obstacles which seemed in the beginning unsurmountable and found elegant solutions of all the problems which confronted me, yet, even at this very day, the majority of experts are still blind to the possibilities which are within easy attainment.

Fig. 11. — Tesla's Four Circuit Tuned System
Contrasted With the Contemporaneous Hertz-wave
System.

My confidence that a signal could be easily flashed around the globe was strengthened thru the discovery of the "rotating brush," a wonderful phenomenon which I have fully described in my address before the Institution of Electrical Engineers, London, in 1892, and which is illustrated in Fig. 9. This is undoubtedly the most delicate wireless detector known, but for a long time it was hard to produce and to maintain in the sensitive state. These difficulties do not exist now and I am looking to valuable applications of this device, particularly in connection with the high-speed photographic method, which I suggested, in wireless, as well as in wire, transmission.

Possibly the most important advances during the following three or four years were my system of concatenated tuned circuits and methods of regulation, now universally adopted. The intimate bearing of these inventions on the development of the wireless art will appear from Fig. 10, which illustrates an arrangement described in my U. S. Patent No. 568178 of September 22, 1896, and corresponding dispositions of wireless apparatus. The captions of the individual diagrams are thought sufficiently explicit to dispense with further comment. I will merely remark that in this early record, in addition to indicating how any number of resonant circuits may be linked and regulated, I have shown the advantage of the proper timing of primary impulses and use of harmonics. In a farcical wireless suit in London, some engineers, reckless of their reputation, have claimed that my circuits were not at all attuned; in fact they asserted that I had looked upon resonance as a sort of wild and untamable beast!

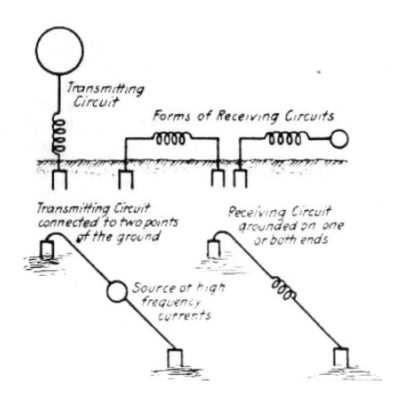

Fig. 12. — Arrangements of Directive Circuits Described In Tesla's U. S. Patent No. 613,809 of November 8. 1898, on "Method of and Apparatus for Controlling Mechanism of Moving Vessels or Vehicles."

It will be of interest to compare my system as first described in a Belgian patent of 1897 with the Hertz-wave system of that period. The significant differences between them will be observed at a glance. The first enables us to transmit economically energy to any distance and is of inestimable value; the latter is capable of a radius of only a few miles and is worthless. In the first there are no spark-gaps and the actions are enormously magnified by resonance. In both transmitter and receiver the currents are transformed and rendered more effective and suitable for the operation of any desired device. Properly constructed, my system is safe against static and other interference and the amount of energy which may be transmitted is *billions of times greater* than with the Hertzian which has none of these virtues, has never been used successfully and of which no trace can be found at present.

Fig. 13. — Diagram Exposing the Fallacy of the Gilding Wave Theory as Propounded In Wireless Text Books.

A well-advertised expert gave out a statement in 1899 that my apparatus did not work and that it would take 200 years before a message would be flashed across the Atlantic and even accepted stolidly my congratulations on a supposed great feat. But subsequent examination of the records showed that my devices were secretly used all the time and ever since I learned of this I have treated these Borgia-Medici methods with the contempt in which they are held by all fair-minded men. The wholesale appropriation of my inventions was, however, not always without a diverting side. As an example to the point I may mention my oscillation transformer operating with an air gap. This was in turn replaced by a carbon arc, quenched gap, an atmosphere of hydrogen, argon or helium, by a mechanical break with oppositely rotating members, a mercury interrupter or some kind of a vacuum bulb and by such *tours de force* as many new "systems" have been produced. I refer to this of course, without the slightest ill-feeling, let us advance by all means. But I cannot help thinking how much better it would have been if the ingenious men, who have originated these "systems," had invented something of their own instead of depending on me altogether.

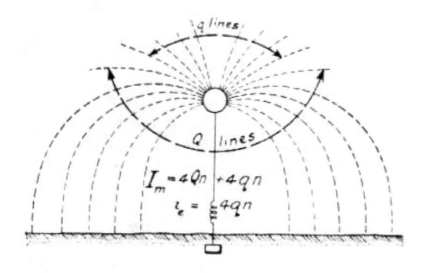

Fig. 14. — Diagram Explaining the Relation Between the Effective and the Measured Current In the Antenna.

Before 1900 two most valuable improvements were made. One of these was my individualized system with transmitters emitting a wave-complex and receivers comprising separate tuned elements coöperatively associated. The underlying principle can be explained in a few words. Suppose that there are n simple vibrations suitable for use in wireless transmission, the probability that any one tune will be struck by an extraneous disturbance is 1 / n. There will then remain $n - 1$ vibrations and the chance that one of these will be excited is 1 / $n - 1$ hence the probability that two tunes would be struck at the same time is 1 / n $(n - 1)$. Similarly, for a combination of three the chance will be 1 / n $(n - 1)$ $(n - 2)$ and so on. It will be readily seen that in this manner any desired degree of safety against the statics or other kind of disturbance can be attained provided the receiving apparatus is so designed that is operation is possible only thru the joint action of all the tuned elements. This was a difficult problem which I have successfully solved so that now *any desired number of simultaneous messages is practicable in the transmission thru the earth as well as thru artificial conductors*.

Fig. 15. — Illustrating One of the General Evidences Against the Space Wave Transmission.

The other invention, of still greater importance, is a peculiar oscillator enabling the transmission of energy without wires in any quantity that may ever be required for industrial use, to any distance, and with very high economy. It was the outcome of years of systematic study and investigation and wonders will be achieved by its means.

The prevailing misconception of the mechanism involved in the wireless transmission has been responsible for various unwarranted announcements which have misled the public and worked harm. By keeping steadily in mind that the transmission thru the earth is in every respect identical to that thru a straight wire, one will gain a clear understanding of the phenomena and will be able to judge correctly the merits of a new scheme. Without wishing to detract from the value of any plan that has been put forward I may say that they are devoid of novelty. So for instance in Fig. 12 arrangements of transmitting and receiving circuits are illustrated, which I have described in my U. S. Patent No. 613809 of November 8, 1898 on a Method of and Apparatus for Controlling Mechanism of Moving Vessels or Vehicles, and which have been recently dished up as original discoveries. In other patents and technical publications I have suggested conductors in the ground as one of the obvious modifications indicated in Fig. 5.

Fig. 16. — Showing Unimportance of Relative Position of Transmitting and Receiving Antennae In Disproval of the Hertz-wave Theory.

For the same reason the statics are still the bane of the wireless. There is about as much virtue in the remedies recently proposed as in hair restorers. *A small and compact apparatus has been produced which does away entirely with this trouble*, at least in plants suitably remodeled.

Nothing is more important in the present phase of development of the wireless art than to dispose of the dominating erroneous ideas. With this object I shall advance a few arguments based on my own observations *which prove that Hertz waves have little to do with the results obtained even at small distances*.

In Fig. 13 a transmitter is shown radiating space waves of considerable frequency. It is generally believed that these waves pass along the earth's surface and thus affect the receivers. I can hardly think of anything more improbable than this "gliding wave" theory and the conception of the "guided wireless" which are contrary to all laws of action and reaction. Why should these disturbances cling to a conductor where they are counteracted by induced currents, when they can propagate in all other directions unimpeded? The fact is that the radiations of the transmitter passing along the earth's surface are soon extinguished, the height of the inactive zone indicated in the diagram, being some function of the wave length, the bulk of the waves traversing freely the atmosphere. Terrestrial phenomena which I have noted conclusively show that there is no *Heaviside layer*, or if it exists, it is of no effect. It certainly would be unfortunate if the human race were thus imprisoned and forever without power to reach out into the depths of space.

Fig. 17. — Illustrating Influence of Obstacle In the Path of Transmission as Evidence Against the Hertz-wave Theory.

The actions at a distance cannot be proportionate to the height of the antenna and the current in the same. I shall endeavor to make this clear by reference to diagram in Fig. 14. The elevated terminal charged to a high potential induces an equal and opposite charge in the earth and there are thus Q lines giving an average current $I = 4Qn$ which circulates locally and is useless except that it adds to the momentum. A relatively small number of lines q however, go off to great distance and to these corresponds a mean current of $ie = 4qn$ to which is due the action at a distance. The total average current in the antenna is thus $Im = 4Qn + 4qn$ and its intensity is no criterion for the performance. The electric efficiency of the antenna is $q / Q+q$ and this is often a very small fraction.

111

Dr. L. W. Austin and Mr. J. L. Hogan have made quantitative measurements which are valuable, but far from supporting the Hertz wave theory they are evidences in disproval of the same, as will be easily perceived by taking the above facts into consideration. Dr. Austin's researches are especially useful and instructive and I regret that I cannot agree with him on this subject. I do not think that if his receiver was affected by Hertz waves he could ever establish such relations as he has found, but he would be likely to reach these results if the Hertz waves were in a large part eliminated. At great distance the space waves and the current waves are of equal energy, the former being merely an accompanying manifestation of the latter in accordance with the fundamental teachings of Maxwell.

Fig. 18. — Showing Effect of Two Hills as Further Proof Against the Hertz-wave Theory.

It occurs to me here to ask the question — why have the Hertz waves been reduced from the original frequencies to those I have advocated for my system, when in so doing the activity of the transmitting apparatus has been reduced a billion fold? I can invite any expert to perform an experiment such as is illustrated in Fig. 15, which shows the classical Hertz oscillator and my grounded transmitting circuit. It is a fact which I have demonstrated that, altho we may have in the Hertz oscillator an activity thousands of times greater, the effect on the receiver is not to be compared to that of the grounded circuit. This shows that *in the transmission from an airplane we are merely working thru a condenser*, the capacity of which is a function of a logarithmic ratio between the length of the conductor and the distance from the ground. The receiver is affected in exactly the same manner as from an ordinary transmitter, the only difference being that there is a certain modification of the action which can be predetermined from the electrical constants. It is not at all difficult to maintain communication between an airplane and a station on the ground, on the contrary, the feat is very easy.

To mention another experiment in support of my view, I may refer to Fig. 16 in which two grounded circuits are shown excited by oscillations of the Hertzian order. It will be found that the antennas can be put out of parallelism without noticeable change in the action on the receiver, this proving that it is due to currents propagated thru the ground and not to space waves.

Fig. 19. — Comparing the Actions of Two Forms of Transmitter as Bearing Out the Fallacy of the Hertz-wave Theory.

Particularly significant are the results obtained in cases illustrated in Figures 17 and 18. In the former an obstacle is shown in the path of the waves but unless the receiver is within the effective *electrostatic* influence of the mountain range, the signals are not appreciably weakened by the presence of the latter, because the currents pass under it and excite the circuit in the same way as if it were attached to an energized wire. If, as in Fig. 18, a second range happens to be beyond the receiver, it could only strengthen the Hertz wave effect by reflection, but as a matter of fact it detracts greatly from the intensity of the received impulses because the electric niveau between the mountains is raised, as I have explained with my <u>lightning protector in the Experimenter of February</u>.

Again in Fig. 19 two transmitting circuits, one grounded directly and the other thru an air gap, are shown. It is a common observation that the former is far more effective, which could not be the case with Hertz radiations. In a like manner if two grounded circuits are observed from day to day the effect is found to increase greatly with the dampness of the ground, and for the same reason also the transmission thru sea-water is more efficient.

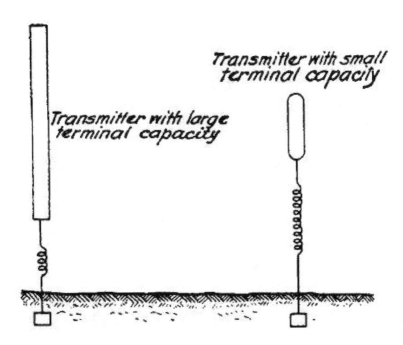

Transmitter with small terminal capacity

Transmitter with large terminal capacity

An illuminating experiment is indicated in Fig. 20 in which two grounded transmitters are shown, one with a large and the other with a small terminal capacity. Suppose that the latter be 1/10 of the former but that it is charged to 10 times the potential and let the frequency of the two circuits and therefore the currents in both antennas be exactly the same. The circuit with the smaller capacity will then have 10 times the energy of the other but the effects on the receiver will be in no wise proportionate.

The same conclusions will be reached by transmitting and receiving circuits with wires buried underground. In each case the actions carefully investigated will be found to be due to *earth currents*. Numerous other proofs might be cited which can be easily verified. So for example *oscillations of low frequency* are ever so much more effective in the transmission which is inconsistent with the prevailing idea. My observations in 1900 and the recent transmissions of signals to very great distances are another emphatic disproval.

The Hertz wave theory of wireless transmission may be kept up for a while, but I do not hesitate to say that in a short time it will be recognized as one of the most remarkable and inexplicable aberrations of the scientific mind which has ever been recorded in history.

TALKING WITH THE PLANETS
by Nikola Tesla

Collier's Weekly

February 9th, 1901

Page number(s): 4-5

Nikola Tesla.

Editor's Note. — *Mr. Nikola Tesla has accomplished some marvelous results in electrical discoveries. Now, with the dawn of the new century, he announces an achievement which will amaze the entire universe, and which eclipses the wildest dream of the most visionary scientist. He has received communication, he asserts, from out the great void of space; a call from the inhabitants of Mars, or Venus, or some other sister planet! And, furthermore, noted scientists, like Sir Norman Lockyer are disposed to agree with Mr. Tesla in his startling deductions.*

Mr. Tesla has not only discovered many important principles, but most of his inventions are in practical use; notably in the harnessing of the Titanic forces of Niagara Falls, and the discovery of a new light by means of a vacuum tube. He has, he declares, solved the problem of telegraphing without wires or artificial conductors of any sort, using the earth as his medium. By means of this principle he expects to be able to send messages under the ocean, and to any distance on the earth's surface. Interplanetary communication has interested him for years, and he sees no reason why we should not soon be within talking distance of Mars or of all worlds in the solar system that may be tenanted by intelligent beings.

At the request of "Collier's Weekly," Mr. Tesla presents herewith a frank statement of what he expects to accomplish and how he hopes to establish communication with the planets.

The idea of communicating with the inhabitants of other worlds is an old one. But for ages it has been regarded merely as a poet's dream, forever unrealizable. And yet, with the invention and perfection of the telescope and the ever-widening knowledge of the heavens, its hold upon our imagination has been increased, and the scientific achievements during the latter part of the nineteenth century, together with the development of the tendency toward the nature ideal of Goethe, have intensified it to such a degree that it seems as if it were destined to become the dominating idea of the century that has just begun. The desire to know something of our neighbors in the immense depths of space does not spring from idle curiosity nor from thirst for knowledge, but from a deeper cause, and it is a feeling firmly rooted in the heart of every human being capable of thinking at all.

Whence, then, does it come? Who knows? Who can assign limits to the subtlety of nature's influences? Perhaps, if we could clearly perceive all the intricate mechanism of the glorious spectacle that is continually unfolding before us, and could, also, trace this desire to its distant origin, we might find it in the sorrowful vibrations of the earth which began when it parted from its celestial parent.

Photographic view of the essential parts of the electrical oscillator used in Mr. Tesla's experiments.

But in this age of reason it is not astonishing to find persons who scoff at the very thought of effecting communication with a planet. First of all, the argument is made that there is only a small probability of other planets being inhabited at all. This argument has never appealed to me. In the solar system, there seem to be only two planets — Venus and Mars — capable of sustaining life such as ours: but this does not mean that there might not be on all of them some other forms of life. Chemical processes may be maintained without the aid of oxygen, and it is still a question whether chemical processes are absolutely necessary to the sustenance of organised beings. My idea is that the development of life must lead to forms of existence that will be possible without nourishment and which will not be shackled by consequent limitations. Why should a living being not be able to obtain all the energy it needs for the performance of its life-functions from the environment, instead of through consumption of food, and transforming, by a complicated process, the energy of chemical combinations into life-sustaining energy?

If there were such beings on one of the planets we should know next to nothing about them. Nor is it necessary to go so far in our assumptions, for we can readily conceive that, in the same degree as the atmosphere diminishes in density, moisture disappears and the planet freezes up, organic life might also undergo corresponding modifications, leading finally to forms which, according to our present ideas of life, are impossible. I will readily admit, of course, that if there should be a sudden catastrophe of any kind all life processes might be arrested; but if the change, no matter how great, should be gradual, and occupied ages, so that the ultimate results could be intelligently foreseen, I cannot but think that reasoning beings would still find means of existence. They would adapt themselves to their constantly changing environment. So I think it quite possible that in a frozen planet, such as our moon is supposed to be, intelligent beings may still dwell, in its interior, if not on its surface.

SIGNALING AT 100,000,000 MILES!

Supplying electrical energy through a single wire without return.

Then it is contended that it is beyond human power and ingenuity to convey signals to the almost inconceivable distances of fifty million or one hundred million miles. This might have been a valid argument formerly. It is not so now. Most of those who are enthusiastic upon the subject of interplanetary communication have reposed their faith in the light-ray as the best possible medium of such communication. True, waves of light, owing to their immense rapidity of succession, can penetrate space more readily than waves less rapid, but a simple consideration will show that by their means an exchange of signals between this earth and its companions in the solar system is, at least now, impossible. By way of illustration, let us suppose that a square mile of the earth's surface — the smallest area that might possibly be within reach of the best telescopic vision of other worlds — were covered with incandescent lamps, packed closely together, so as to form, when illuminated, a continuous sheet of light. It would require not less than one hundred million horse-power to light this area of lamps, and this is many times the amount of motive power now in the service of man throughout the world.

But with the novel means, proposed by myself, I can readily demonstrate that, with an expenditure not exceeding two thousand horse-power, signals can be transmitted to a planet such as Mars with as much exactness and certitude as we now send messages by wire from New York to Philadelphia. These means are the result of long-continued experiment and gradual improvement.

Some ten years ago, I recognised the fact that to convey electric currents to a distance it was not at all necessary to employ a return wire, but that any amount of energy might be transmitted by using a single wire. I illustrated this principle by numerous experiments, which, at that time, excited considerable attention among scientific men.

This being practically demonstrated, my next step was to use the earth itself as the medium for conducting the currents, thus dispensing with wires and all other artificial conductors. So I was led to the development of a system of energy transmission and of telegraphy without the use of wires, which I described in 1893. The difficulties I encountered at first in the transmission of currents through the earth were very great. At that time I had at hand only ordinary apparatus, which I found to be ineffective, and I concentrated my attention immediately upon perfecting machines for this special purpose. This work consumed a number of years, but I finally vanquished all difficulties and succeeded in producing a machine which, to explain its operation in plain language, resembled a pump in its action, drawing electricity from the earth and driving it back into the same at an enormous rate, thus creating ripples or disturbances which, spreading through the earth as through a wire, could be detected at great distances by carefully attuned receiving circuits. In this manner I was able to transmit to a distance, not only feeble effects for the purposes of signalling, but considerable amounts of energy, and later discoveries I made convinced me that I shall ultimately succeed in conveying power without wires, for industrial purposes, with high economy, and to any distance, however great.

EXPERIMENTS IN COLORADO

An electrical oscillator delivering energy at a rate of 75,000 horse-power.

To develop these inventions further, I went to Colorado in 1899, where I continued my investigations along these and other lines, one of which, in particular, I now consider of even greater importance than the transmission of power without wires. I constructed a laboratory in the neighborhood of Pike's Peak. The conditions in the pure air of the Colorado Mountains proved extremely favorable for my experiments, and the results were most gratifying to me. I found that I could not only accomplish more work, physically and mentally, than I could in New

York, but that electrical effects and changes were more readily and distinctly perceived. A few years ago it was virtually impossible to produce electrical sparks twenty or thirty foot long; but I produced some more than one hundred feet in length, and this without difficulty. The rates of electrical movement involved in strong induction apparatus had measured but a few hundred horse-power, and I produced electrical movements of rates of one hundred and ten thousand horse-power. Prior to this, only insignificant electrical pressures were obtained, while I have reached fifty million volts.

The accompanying illustrations, with their descriptive titles, taken from an article I wrote for the "Century Magazine," may serve to convey an idea of the results I obtained in the directions indicated.

Many persons in my own profession have wondered at them and have asked what I am trying to do. But the time is not far away now when the practical results of my labors will be placed before the world and their influence felt everywhere. One of the immediate consequences will be the transmission of messages without wires, over sea or land, to an immense distance. I have already demonstrated, by crucial tests, the practicability of signalling by my system from one to any other point of the globe, no matter how remote, and I shall soon convert the disbelievers.

I have every reason for congratulating myself that throughout these experiments, many of which were exceedingly delicate and hazardous, neither myself nor any of my assistants received any injury. When working with these powerful electrical oscillations the most extraordinary phenomena take place at times. Owing to some interference of the oscillations, veritable balls of fire are apt to leap out to a great distance, and if any one were within or near their paths, he would be instantly destroyed. A machine such as I have used could easily kill, in an instant, three hundred thousand persons. I observed that the strain upon my assistants was telling, and some of them could not endure the extreme tension of the nerves. But these perils are now entirely overcome, and the operation of such apparatus, however powerful, involves no risk whatever.

As I was improving my machines for the production of intense electrical actions, I was also perfecting the means for observing feeble effects. One of the most interesting results, and also one of great practical importance, was the development of certain contrivances for indicating at a distance of many hundred miles an approaching storm, its direction, speed and distance travelled. These appliances are likely to be valuable in future meteorological observations and surveying, and will lend themselves particularly to many naval uses.

It was in carrying on this work that for the first time I discovered those mysterious effects which have elicited such unusual interest. I had perfected the apparatus referred to so far that from my laboratory in the Colorado mountains I could feel the pulse of the globe, as it were, noting every electrical change that occurred within a radius of eleven hundred miles.

TERRIFIED BY SUCCESS

Transmitting electrical energy through the earth without wire.

I can never forget the first sensations I experienced when it dawned upon me that I had observed something possibly of incalculable consequences to mankind. I felt as though I were present at the birth of a new knowledge or the revelation of a great truth. Even now, at times, I can vividly recall the incident, and see my apparatus as though it were actually before me. My first observations positively terrified me, as there was present in them something mysterious, not to say supernatural, and I was alone in my laboratory at night; but at that time the idea of these disturbances being intelligently controlled signals did not yet present itself to me.

The changes I noted were taking place periodically, and with such a clear suggestion of number and order that they were not traceable to any cause then known to me. I was familiar, of course, with such electrical disturbances as are produced by the sun, Aurora Borealis and earth currents, and I was as sure as I could be of any fact that these variations were due to none of these causes. The nature of my experiments precluded the possibility of the changes being produced by atmospheric disturbances, as has been rashly asserted by some. It was some time afterward when the thought flashed upon my mind that the disturbances I had observed might be due to an intelligent control. Although I could not decipher their meaning, it was impossible for me to think of them as having been entirely accidental. The feeling is constantly growing on me that I had been the first to hear the greeting of one planet to another. A purpose was behind these electrical signals; and it was with this conviction that I announced to the Red Cross Society, when it asked me to indicate one of the great possible achievements of the next hundred years, that it would probably be the confirmation and interpretation of this planetary challenge to us.

Since my return to New York more urgent work has consumed all my attention; but I have never ceased to think of those experiences and of the observations made in Colorado. I am constantly endeavoring to improve and perfect my apparatus, and just as soon as practicable I shall again take up the thread of my investigations at the point where I have been forced to lay it down for a time.

COMMUNICATING WITH THE MARTIANS

At the present stage of progress, there would be no insurmountable obstacle in constructing a machine capable of conveying a message to Mars, nor would there be any great difficulty in recording signals transmitted to us by the inhabitants of that planet, if they be skilled electricians. Communication once established, even in the simplest way, as by a mere interchange of numbers, the progress toward more intelligible communication would be rapid. Absolute certitude as to the receipt and interchange of messages would be reached as soon as we could respond with the number "four," say, in reply to the signal "one, two, three." The Martians, or the inhabitants of whatever planet had signalled to us, would understand at once that we had caught their message across the gulf of space and had sent back a response. To convey a knowledge of form by such means is, while very difficult, not impossible, and I have already found a way of doing it.

Experiment to illustrate the capacity of the oscillator for producing electrical explosions of great power.

What a tremendous stir this would make in the world! How soon will it come? For that it will some time be accomplished must be clear to every thoughtful being.

Something, at least, science has gained. But I hope that it will also be demonstrated soon that in my experiments in the West I was not merely beholding a vision, but had caught sight of a great and profound truth.

THE PROBLEM OF INCREASING HUMAN ENERGY

by Nikola Tesla
Century Magazine, The

June, 1900
Page number(s):

175-211

The Problem of Increasing Human Energy
With Special Reference to the Harnessing of the Sun's Energy.
by Nikola Tesla

Illustrated by The Writer's Electrical Experiments, Now First Published

THE ONWARD MOVEMENT OF MAN — THE ENERGY OF THE MOVEMENT — THE THREE WAYS OF INCREASING HUMAN ENERGY.

Of all the endless variety of phenomena which nature presents to our senses, there is none that fills our minds with greater wonder than that inconceivably complex movement which, in its entirety, we designate as human life. Its mysterious origin is veiled in the forever impenetrable mist of the past, its character is rendered incomprehensible by its infinite intricacy, and its destination is hidden in the unfathomable depths of the future. Whence does it come? What is it? Whither does it tend? are the great questions which the sages of all times have endeavored to answer.

Modern science says: The sun is the past, the earth is the present, the moon is the future. From an incandescent mass we have originated, and into a frozen mass we shall turn. Merciless is the law of nature, and rapidly and irresistibly we are drawn to our doom. Lord Kelvin, in his profound meditations, allows us only a short span of life, something like six million years, after which time the sun's bright light will have ceased to shine, and its life-giving heat will have ebbed away, and our own earth will be a lump of ice, hurrying on through the eternal night. But do not let us despair. There will still be left on it a glimmering spark of life, and there will be a chance to kindle a new fire on some distant star. This wonderful possibility seems, indeed, to exist, judging from Professor Dewar's beautiful experiments with liquid air, which show that germs of organic life are not destroyed by cold, no matter how intense; consequently they may be transmitted through the interstellar space. Meanwhile the cheering lights of science and art, ever increasing in intensity, illuminate our path, and marvels they disclose, and the enjoyments they offer, make us measurably forgetful of the gloomy future.

Though we may never be able to comprehend human life, we know certainly that it is a movement, of whatever nature it be. The existence of movement unavoidably implies a body which is being moved and a force which is moving it. Hence, wherever there is life, there is a mass moved by a force. All mass possesses inertia, all force tends to persist. Owing to this universal property and condition, a body, be it at rest or in motion, tends to remain in the same state, and a force, manifesting itself anywhere and through whatever cause, produces an equivalent opposing force, and as an absolute necessity of this it follows that every movement in nature must be rhythmical. Long ago this simple truth was clearly pointed out by Herbert Spencer, who arrived at it through a somewhat different process of reasoning. It is borne out in everything we perceive — in the movement of a planet, in the surging and ebbing of the tide, in the reverberations of the air, the swinging of a pendulum, the oscillations of an electric current, and in the infinitely varied phenomena of organic life. Does not the whole of human life attest to it? Birth, growth, old age, and death of an individual, family, race, or nation, what is it all but a rhythm? All life-manifestation, then, even in its most intricate form, as exemplified in man, however involved and inscrutable, is only a movement, to which the same general laws of movement which govern throughout the physical universe must be applicable.

Fig. 1. — Burning the Nitrogen of the Atmosphere.

Note to Fig. 1. — This result is produced by the discharge of an electrical oscillator giving twelve

135

million volts. The electrical pressure, alternating one hundred thousand times per second, excites the normally inert nitrogen, causing it to combine with the oxygen. The flame-like discharge shown in the photograph measures sixty-five feet across.

When we speak of man, we have a conception of humanity as a whole, and before applying scientific methods to the investigation of his movement, we must accept this as a physical fact. But can anyone doubt to-day that all the millions of individuals and all the innumerable types and characters constitute an entity, a unit? Though free to think and act, we are held together, like the stars in the firmament, with ties inseparable. These ties cannot be seen, but we can feel them. I cut myself in the finger, and it pains me: this finger is a part of me. I see a friend hurt, and it hurts me, too: my friend and I are one. And now I see stricken down an enemy, a lump of matter which, of all the lumps of matter in the universe, I care least for, and it still grieves me. Does this not prove that each of us is only part of a whole?

For ages this idea has been proclaimed in the consummately wise teachings of religion, probably not alone as a means of insuring peace and harmony among men, but as a deeply founded truth. The Buddhist expresses it in one way, the Christian in another, but both say the same: We are all one. Metaphysical proofs are, however, not the only ones which we are able to bring forth in support of this idea. Science, too, recognizes this connectedness of separate individuals, though not quite in the same sense as it admits that the suns, planets, and moons of a constellation are one body, and there can be no doubt that it will be experimentally confirmed in times to come, when our means and methods for investigating psychical and other states and phenomena shall have been brought to great perfection. Still more: this one human being lives on and on. The individual is ephemeral, races and nations come and pass away, but man remains. Therein lies the profound difference between the individual and the whole. Therein, too, is to be found the partial explanation of many of those marvelous phenomena of heredity which are the result of countless centuries of feeble but persistent influence.

Conceive, then, man as a mass urged on by a force. Though this movement is not of a translatory character, implying change of place, yet the general laws of mechanical movement are applicable to it, and the energy associated with this mass can be measured, in accordance with well-known principles, by half the product of the mass with the square of a certain velocity. So, for instance, a cannon-ball which is at rest possesses a certain amount of energy in the form of heat, which we measure in a similar way. We imagine the ball to consist of innumerable minute particles, called atoms or molecules, which vibrate or whirl around one another. We determine their masses and velocities, and from them the energy of each of these minute systems, and adding them all together, we get an idea of the total heat-energy contained in the ball, which is only seemingly at rest. In this purely theoretical estimate this energy may then be calculated by multiplying half of the total mass — that is, half of the sum of all the small masses — with the square of a velocity which is determined from the velocities of the separate particles. In like manner we may conceive of human energy being measured by half the human mass multiplied with the square of a velocity which we are not yet able to compute. But our deficiency in this knowledge will not vitiate the truth of the deductions I shall draw, which rest on the firm basis that the same laws of mass and force govern throughout nature.

Man, however, is not an ordinary mass, consisting of spinning atoms and molecules, and containing merely heat-energy. He is a mass possessed of certain higher qualities by reason of the creative principle of life with which he is endowed. His mass, as the water in an ocean wave, is being continuously exchanged, new taking the place of the old. Not only this, but he grows, propagates, and dies, thus altering his mass independently, both in bulk and density. What is most wonderful of all, he is capable of increasing or diminishing his velocity of movement by the mysterious power he possesses of appropriating more or less energy from other substance, and turning it into motive energy. But in any given moment we may ignore these slow changes and assume that human energy is measured by half the product of man's mass with the square of a certain hypothetical velocity. However we may compute this velocity, and whatever we may take as the standard of its measure, we must, in harmony with this conception, come to the conclusion that the great problem of science is, and always will be, to increase the energy thus defined. Many years ago, stimulated by the perusal of that deeply interesting work, Draper's "History of the Intellectual Development of Europe," depicting so vividly human movement, I recognized that to solve this eternal problem must ever be the chief task of the man of science. Some results of my own efforts to this end I shall endeavor briefly to describe here.

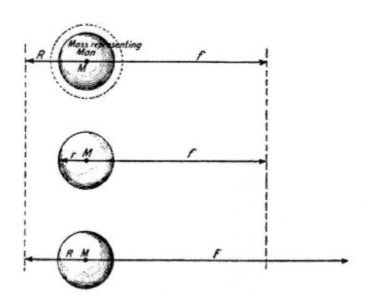

Diagram a. — The Three Ways of Increasing Human Energy.

Let, then, in diagram *a*, *M* represent the mass of man. This mass is impelled in one direction by a force *f*, which is resisted by another partly frictional and partly negative force *R*, acting in a direction exactly opposite, and retarding the movement of the mass. Such an antagonistic force is present in every movement and must be taken into consideration. The difference between these two forces is the effective force which imparts a velocity *V* to the mass *M* in the direction of the arrow on the line representing the force *f*. In accordance with the preceding, the human energy will then be given by the product $\frac{1}{2} MV^2 = \frac{1}{2} MV \times V$, in which *M* is the total mass of man in the ordinary interpretation of the term "mass," and *V* is a certain hypothetical velocity, which, in the present state of science, we are unable exactly to define and determine. To increase the human energy is, therefore, equivalent to increasing this product, and there are, as will readily be seen, only three ways

139

possible to attain this result, which are illustrated in the above diagram. The first way, shown in the top figure, is to increase the mass (as indicated by the dotted circle), leaving the two opposing forces the same. The second way is to reduce the retarding force R to a smaller value r, leaving the mass and the impelling force the same, as diagrammatically shown in the middle figure. The third way, which is illustrated in the last figure, is to increase the impelling force f to a higher value F, while the mass and the retarding force R remain unaltered. Evidently fixed limits exist as regards increase of mass and reduction of retarding force, but the impelling force can be increased indefinitely. Each of these three possible solutions presents a different aspect of the main problem of increasing human energy, which is thus divided into three distinct problems, to be successively considered.

THE FIRST PROBLEM: HOW TO INCREASE THE HUMAN MASS — THE BURNING OF ATMOSPHERIC NITROGEN.

Viewed generally, there are obviously two ways of increasing the mass of mankind: first, by aiding and maintaining those forces and conditions which tend to increase it; and, second, by opposing and reducing those which tend to diminish it. The mass will be increased by careful attention to health, by substantial food, by moderation, by regularity of habits, by promotion of marriage, by conscientious attention to the children, and, generally stated, by the observance of all the many precepts and laws of religion and hygiene. But in adding new mass to the old, three cases again present themselves. Either the mass added is of the same velocity as the old, or it is of a smaller or of a higher velocity. To gain an idea of the relative importance of these cases, imagine a train composed of, say, one hundred locomotives running on a track, and suppose that, to increase the energy of the moving mass, four more locomotives are added to the train. If these four move at the same velocity at which the train is going, the total energy will be increased four per cent.; if they are moving at only one half of that velocity, the increase will amount to only one per cent.; if they are moving at twice that velocity, the increase of energy will be sixteen per cent. This simple illustration shows that it is of the greatest importance to add mass of a higher velocity. Stated more to the point, if, for example, the children be of the same degree of enlightenment as the parents, — that is, mass of the "same velocity," — the energy will simply increase proportionately to the number added. If they are less intelligent or advanced, or mass of "smaller velocity," there will be a very slight gain in the energy; but if they are further advanced, or mass of "higher velocity," then the new generation will add very considerably to the sum total of human energy. Any addition of mass of "smaller velocity," beyond that indispensable amount required by the law expressed in the proverb, "Mens sana in corpore sano," should be strenuously opposed. For instance, the mere development of muscle, as aimed at in some of our colleges, I consider equivalent to adding mass of "smaller velocity," and I would not commend it, although my views were different when I was a student myself. Moderate exercise, insuring the right balance between mind and body, and the highest efficiency of performance, is, of course, a prime requirement. The above example shows that the most important result to be attained is the education, or the increase of the "velocity," of the mass newly added.

Conversely, it scarcely need be stated that everything that is against the teachings of religion and the laws of hygiene is tending to decrease the mass. Whisky, wine, tea, coffee, tobacco, and other such stimulants are responsible for the shortening of the lives of many, and ought to be used with moderation. But I do not think that rigorous measures of suppression of habits followed through many generations are commendable. It is wiser to preach moderation than abstinence. We have become accustomed to these stimulants, and if such reforms are to be effected, they must be slow and gradual. Those who are devoting their energies to such ends could make themselves far more useful by turning their efforts in other directions, as, for instance, toward providing pure water.

For every person who perishes from the effects of a stimulant, at least a thousand die from the consequences of drinking impure water. This precious fluid, which daily infuses new life into us, is likewise the chief vehicle through which disease and death enter our bodies. The germs of destruction it conveys are enemies all the more terrible as they perform their fatal work unperceived. They seal our doom while we live and enjoy. The majority of people are so ignorant or careless in drinking water, and the consequences of this are so disastrous, that a philanthropist can scarcely use his efforts better than by endeavoring to enlighten those who are thus injuring themselves. By systematic purification and sterilization of the drinking-water the human mass would be very considerably increased. It should be made a rigid rule — which might be enforced by law — to boil or to sterilize otherwise the drinking-water in every household and public place. The mere filtering does not afford sufficient security against infection. All ice for internal uses should be artificially prepared from water thoroughly sterilized. The importance of eliminating germs of disease from the city water is generally recognized, but little is being done to improve the existing conditions, as no satisfactory method of sterilizing great quantities of water has yet been brought forward. By improved electrical appliances we are now enabled to produce ozone cheaply and in large amounts, and this ideal disinfectant seems to offer a happy solution of the important question.

Gambling, business rush, and excitement, particularly on the exchanges, are causes of much mass-reduction, all the more so because the individuals concerned represent units of higher value. Incapacity of observing the first symptoms of an illness, and careless neglect of the same, are important factors of mortality. In noting carefully every new sign of approaching danger, and making conscientiously every possible effort to avert it, we are not only following wise laws of hygiene in the interest of our well-being and the success of our labors, but we are also complying with a higher moral duty. Every one should consider his body as a priceless gift from one whom he loves above all, as a marvelous work of art, of undescribable beauty and mastery beyond human conception, and so delicate and frail that a word, a breath, a look, nay, a thought, may injure it. Uncleanliness, which breeds disease and death, is not only a self-destructive but highly immoral habit. In keeping our bodies free from infection, healthful, and pure, we are expressing our reverence for the high principle with which they are endowed. He who follows the precepts of hygiene in this spirit is proving himself, so far, truly religious. Laxity of morals is a terrible evil, which poisons both mind and body, and which is responsible for a great reduction of the human mass in some countries. Many of the present customs and tendencies are productive of similar hurtful results. For example, the society life, modern education and pursuits of women, tending to draw them away from their household duties and make men out of them, must needs detract from the elevating ideal they represent, diminish the artistic creative power, and cause sterility and a general weakening of the race. A thousand other evils might be mentioned, but all put together, in their bearing upon the problem under discussion, they would not equal a single one, the want of food, brought on by poverty, destitution, and famine. Millions of individuals die yearly for want of food, thus keeping down the mass. Even in our enlightened communities, and not withstanding the many charitable efforts, this is still, in all probability, the chief evil. I do not mean here absolute want of food, but want of healthful nutriment.

How to provide good and plentiful food is, therefore, a most important question of the day. On the general principles the raising of cattle as a means of providing food is objectionable, because, in the sense interpreted above, it must undoubtedly tend to the addition of mass of a "smaller velocity." It is certainly preferable to raise vegetables, and I think, therefore, that vegetarianism is a commendable departure from the established barbarous habit. That we can subsist on plant food and perform our work even to advantage is not a theory, but a well-demonstrated fact. Many races living almost exclusively on vegetables are of superior physique and strength. There is no doubt that some plant food, such as oatmeal, is more economical than meat, and superior to it in regard to both mechanical and mental performance. Such food, moreover, taxes our digestive organs decidedly less, and, in making us more contented and sociable, produces an amount of good difficult to estimate. In view of these facts every effort should be made to stop the wanton and cruel slaughter of animals, which must be destructive to our morals. To free ourselves from animal instincts and appetites, which keep us down, we should begin at the very root from which we spring: we should effect a radical reform in the character of the food.

There seems to be no *philosophical* necessity for food. We can conceive of organized beings living without nourishment, and deriving all the energy they need for the performance of their life-functions from the ambient medium. In a crystal we have the clear evidence of the existence of a formative life-principle, and though we cannot understand the life of a crystal, it is none the less a living being. There may be, besides crystals, other such individualized, material systems of beings, perhaps of gaseous constitution, or composed of substance still more tenuous. In view of this possibility, — nay, probability, — we cannot apodictically deny the existence of organized beings on a planet merely because the conditions on the same are unsuitable for the existence of life as we conceive it. We cannot even, with positive assurance, assert that some of them might not be present here, in this our world, in the very midst of us, for their constitution and life-manifestation may be such that we are unable to perceive them.

The production of artificial food as a means for causing an increase of the human mass naturally suggests itself, but a direct attempt of this kind to provide nourishment does not appear to me rational, at least not for the present. Whether we could thrive on such food is very doubtful. We are the result of ages of continuous adaptation, and we cannot radically change without unforeseen and, in all probability, disastrous consequences. So uncertain an experiment should not be tried. By far the best way, it seems to me, to meet the ravages of the evil, would be to find ways of increasing the productivity of the soil. With this object the preservation of forests is of an importance which cannot be overestimated, and in this connection, also, the utilization of water-power for purposes of electrical transmission, dispensing in many ways with the necessity of burning wood, and tending thereby to forest preservation, is to be strongly advocated. But there are limits in the improvement to be effected in this and similar ways.

To increase materially the productivity of the soil, it must be more effectively fertilized by artificial means. The question of food-production resolves itself, then, into the question how best to fertilize the soil. What it is that made the soil is still a mystery. To explain its origin is probably equivalent to explaining the origin of life itself. The rocks, disintegrated by moisture and heat and wind and weather, were in themselves not capable of maintaining life. Some unexplained condition arose, and some new principle came into effect, and the first layer capable of sustaining low organisms, like mosses, was formed. These, by their life and death, added more of the life-sustaining quality to the soil, and higher organisms could then subsist, and so on and on, until at last highly developed plant and animal life could flourish. But though the theories are, even now, not in agreement as to how fertilization is effected, it is a fact, only too well ascertained, that the soil cannot indefinitely sustain life, and some way must be found to supply it with the substances which have been abstracted from it by the plants. The chief and most valuable among these substances are compounds of nitrogen, and the cheap production of these is, therefore, the key for the solution of the all-important food problem. Our atmosphere contains an inexhaustible amount of nitrogen, and could we but oxidize it and produce these compounds, an incalculable benefit for mankind would follow.

Long ago this idea took a powerful hold on the imagination of scientific men, but an efficient means for accomplishing this result could not be devised.

The problem was rendered extremely difficult by the extraordinary inertness of the nitrogen, which refuses to combine even with oxygen. But here electricity comes to our aid: the dormant affinities of the element are awakened by an electric current of the proper quality. As a lump of coal which has been in contact with oxygen for centuries without burning will combine with it when once ignited, so nitrogen, excited by electricity, will burn. I did not succeed, however, in producing electrical discharges exciting very effectively the atmospheric nitrogen until a comparatively recent date, although I showed, in May, 1891, in a scientific lecture, a novel form of discharge or electrical flame named "St. Elmo's hotfire," which, besides being capable of generating ozone in abundance, also possessed, as I pointed out on that occasion, distinctly the quality of exciting chemical affinities. This discharge or flame was then only three or four inches long, its chemical action was likewise very feeble, and consequently the process of oxidation of nitrogen was wasteful. How to intensify this action was the question. Evidently electric currents of a peculiar kind had to be produced in order to render the process of nitrogen combustion more efficient.

The first advance was made in ascertaining that the chemical activity of the discharge was very considerably increased by using currents of extremely high frequency or rate of vibration. This was an important improvement, but practical considerations soon set a definite limit to the progress in this direction. Next, the effects of the electrical pressure of the current impulses, of their wave-form and other characteristic features, were investigated. Then the influence of the atmospheric pressure and temperature and of the presence of water and other bodies was studied, and thus the best conditions for causing the most intense chemical action of the discharge and securing the highest efficiency of the process were gradually ascertained. Naturally, the improvements were not quick in coming; still, little by little, I advanced. The flame grew larger and larger, and its oxidizing action grew more intense. From an insignificant brush-discharge a few inches long it developed into a marvelous electrical phenomenon, a roaring blaze, devouring the nitrogen of the atmosphere and measuring sixty or seventy feet across. Thus slowly, almost imperceptibly, possibility became accomplishment. All is not yet done, by any means, but to what a degree my efforts have been rewarded an idea may be gained from an inspection of Fig. 1 (p. 176), which, with its title, is self-explanatory. The flame-like discharge visible is produced by the intense electrical oscillations which pass through the coil shown, and violently agitate the electrified molecules of the air. By this means a strong affinity is created between the two normally indifferent constituents of the atmosphere, and they combine readily, even if no further provision is made for intensifying the chemical action of the discharge. In the manufacture of nitrogen compounds by this method, of course, every possible means bearing upon the intensity of this action and the efficiency of the process will be taken advantage of, and, besides, special arrangements will be provided for the fixation of the compounds formed, as they are generally unstable, the nitrogen becoming again inert after a little lapse of time. Steam is a simple and effective means for fixing permanently the compounds. The result illustrated makes it practicable to oxidize the atmospheric nitrogen in unlimited quantities, merely by the use of cheap mechanical power and simple electrical apparatus. In this manner many compounds of nitrogen may be manufactured all over the world, at a small cost, and in any desired amount, and by means of these compounds the soil can be fertilized and its productiveness indefinitely increased. An abundance of cheap and healthful food, not artificial, but such as we are accustomed to, may thus be obtained. This new and inexhaustible source of food-supply will be of incalculable benefit to mankind, for it will enormously contribute to the increase of the human mass, and thus add immensely to human energy. Soon, I hope, the world will see the beginning of an industry which, in time to come, will, I believe, be in importance next to that of iron.

THE SECOND PROBLEM: HOW TO REDUCE THE FORCE RETARDING THE HUMAN MASS — THE ART OF TELAUTOMATICS.

As before stated, the force which retards the onward movement of man is partly frictional and partly negative. To illustrate this distinction I may name, for example, ignorance, stupidity, and imbecility as some of the purely frictional forces, or resistances devoid of any directive tendency. On the other hand, visionariness, insanity, self-destructive tendency, religious fanaticism, and the like, are all forces of a negative character, acting in definite directions. To reduce or entirely to overcome these dissimilar retarding forces, radically different methods must be employed. One knows, for instance, what a fanatic may do, and one can take preventive measures, can enlighten, convince, and possibly direct him, turn his vice into virtue; but one does not know, and never can know, what a brute or an imbecile may do, and one must deal with him as with a mass, inert, without mind, let loose by the mad elements. A negative force always implies some quality, not infrequently a high one, though badly directed, which it is possible to turn to good advantage; but a directionless, frictional force involves unavoidable loss. Evidently, then, the first and general answer to the above question is:

turn all negative force in the right direction and reduce all frictional force.

There can be no doubt that, of all the frictional resistances, the one that most retards human movement is ignorance. Not without reason said that man of wisdom, Buddha: "Ignorance is the greatest evil in the world." The friction which results from ignorance, and which is greatly increased owing to the numerous languages and nationalities, can be reduced only by the spread of knowledge and the unification of the heterogeneous elements of humanity. No effort could be better spent. But however ignorance may have retarded the onward movement of man in times past, it is certain that, nowadays, negative forces have become of greater importance. Among these there is one of far greater moment than any other. It is called organized warfare. When we consider the millions of individuals, often the ablest in mind and body, the flower of humanity, who are compelled to a life of inactivity and unproductiveness, the immense sums of money daily required for the maintenance of armies and war apparatus, representing ever so much of human energy, all the effort uselessly spent in the production of arms and implements of destruction, the loss of life and the fostering of a barbarous spirit, we are appalled at the inestimable loss to mankind which the existence of these deplorable conditions must involve. What can we do to combat best this great evil?

Law and order absolutely require the maintenance of organized force. No community can exist and prosper without rigid discipline. Every country must be able to defend itself, should the necessity arise. The conditions of to-day are not the result of yesterday, and a radical change cannot be effected to-morrow. If the nations would at once disarm, it is more than likely that a state of things worse than war itself would follow. Universal peace is a beautiful dream, but not at once realizable. We have seen recently that even the nobel effort of the man invested with the greatest worldly power has been virtually without effect. And no wonder, for the establishment of universal peace is, for the time being, a physical impossibility. War is a negative force, and cannot be turned in a positive direction without passing through the intermediate phases. It is the problem of making a wheel, rotating one way, turn in the opposite direction without slowing it down, stopping it, and speeding it up again the other way.

It has been argued that the perfection of guns of great destructive power will stop warfare. So I myself thought for a long time, but now I believe this to be a profound mistake. Such developments will greatly modify, but not arrest it. On the contrary, I think that every new arm that is invented, every new departure that is made in this direction, merely invites new talent and skill, engages new effort, offers new incentive, and so only gives a fresh impetus to further development. Think of the discovery of gun-powder. Can we conceive of any more radical departure than was effected by this innovation? Let us imagine ourselves living in that period: would we not have thought then that warfare was at an end, when the armor of the knight became an object of ridicule, when bodily strength and skill, meaning so much before, became of comparatively little value? Yet gunpowder did not stop warfare; quite the opposite — it acted as a most powerful incentive. Nor do I believe that warfare can ever be arrested by any scientific or ideal development, so long as similar conditions to those prevailing now exist, because war has itself become a science, and because war involves some of the most sacred sentiments of which man is capable. In fact, it is doubtful whether men who would not be ready to fight for a high principle would be good for anything at all. It is not the mind which makes man, nor is it the body; it is mind and body. Our virtues and our failings are inseparable, like force and matter. When they separate, man is no more.

Another argument, which carries considerable force, is frequently made, namely, that war must soon become impossible be cause the means of defense are outstripping the means of attack. This is only in accordance with a fundamental law which may be expressed by the statement that it is easier to destroy than to build. This law defines human capacities and human conditions. Were these such that it would be easier to build than to destroy, man would go on unresisted, creating and accumulating without limit. Such conditions are not of this earth. A being which could do this would not be a man; it might be a god. Defense will always have the advantage over attack, but this alone, it seems to me, can never stop war. By the use of new principles of defense we can render harbors impregnable against attack, but we cannot by such means prevent two war-ships meeting in battle on the high sea. And then, if we follow this idea to its ultimate development, we are led to the conclusion that it would be better for mankind if attack and defense were just oppositely related; for if every country, even the smallest, could surround itself with a wall absolutely impenetrable, and could defy the rest of the world, a state of things would surely be brought on which would be extremely unfavorable to human progress. It is by abolishing all the barriers which separate nations and countries that civilization is best furthered.

Again, it is contended by some that the advent of the flying-machine must bring on universal peace. This, too, I believe to be an entirely erroneous view. The flying-machine is certainly coming, and very soon, but the conditions will remain the same as before. In fact, I see no reason why a ruling power, like Great Britain, might not govern the air as well as the sea. Without wishing to put myself on record as a prophet, I do not hesitate to say that the next years will see the establishment of an "air-power," and its center may be not far from New York. But, for all that, men will fight on merrily.

The ideal development of the war principle would ultimately lead to the transformation of the whole energy of war into purely potential, explosive energy, like that of an electrical condenser. In this form the war-energy could be maintained without effort; it would need to be much smaller in amount, while incomparably more effective.

As regards the security of a country against foreign invasion, it is interesting to note that it depends only on the relative, and not the absolute, number of the individuals or magnitude of the forces, and that, if every country should reduce the war-force in the same ratio, the security would remain unaltered. An international agreement with the object of reducing to a minimum the war-force which, in view of the present still imperfect education of the masses, is absolutely indispensable, would, therefore, seem to be the first rational step to take toward diminishing the force retarding human movement.

Fortunately, the existing conditions cannot continue indefinitely, for a new element is beginning to assert itself. A change for the better is imminent, and I shall now endeavor to show what, according to my ideas, will be the first advance toward the establishment of peaceful relations between nations, and by what means it will eventually be accomplished.

Let us go back to the early beginning, when the law of the stronger was the only law. The light of reason was not yet kindled, and the weak was entirely at the mercy of the strong. The weak individual then began to learn how to defend himself. He made use of a club, stone, spear, sling, or bow and arrow, and in the course of time, instead of physical strength, intelligence became the chief deciding factor in the battle. The wild character was gradually softened by the awakening of noble sentiments, and so, imperceptibly, after ages of continued progress, we have come from the brutal fight of the unreasoning animal to what we call the "civilized warfare" of to-day, in which the combatants shake hands, talk in a friendly way, and smoke cigars in the entr'actes, ready to engage again in deadly conflict at a signal. Let pessimists say what they like, here is an absolute evidence of great and gratifying advance.

But now, what is the next phase in this evolution? Not peace as yet, by any means. The next change which should naturally follow from modern developments should be the continuous diminution of the number of individuals engaged in battle. The apparatus will be one of specifically great power, but only a few individuals will be required to operate it. This evolution will bring more and more into prominence a machine or mechanism with the fewest individuals as an element of warfare, and the absolutely unavoidable consequence of this will be the abandonment of large, clumsy, slowly moving, and unmanageable units. Greatest possible speed and maximum rate of energy-delivery by the war apparatus will be the main object. The loss of life will become smaller and smaller, and finally, the number of the individuals continuously diminishing, merely machines will meet in a contest without blood-shed, the nations being simply interested, ambitious spectators. When this happy condition is realized, peace will be assured. But, no matter to what degree of perfection rapid-fire guns, high-power cannon, explosive projectiles, torpedo-boats, or other implements of war may be brought, no matter how destructive they may be made, that condition can never be reached through any such development. All such implements require men for their operation; men are indispensable parts of the machinery. Their object is to kill and to destroy. Their power resides in their capacity for doing evil. So long as men meet in battle, there will be bloodshed. Bloodshed will ever keep up barbarous passion. To break this fierce spirit, a radical departure must be made, an entirely new principle must be introduced, something that never existed before in warfare — a principle which will forcibly, unavoidably, turn the battle into a mere spectacle, a play, a contest without loss of blood. To bring on this result men must be dispensed with: machine must fight machine. But how accomplish that which seems impossible? The answer is simple enough: produce a machine capable of acting as though it were part of a human being — no mere mechanical contrivance, comprising levers, screws, wheels, clutches, and nothing more, but a machine embodying a higher principle, which will enable it to perform its duties as though it had intelligence, experience, judgment, a mind! This conclusion is the result of my thoughts and observations which have extended through virtually my whole life, and I shall now briefly describe how I came to accomplish that which at first seemed an unrealizable dream.

A long time ago, when I was a boy, I was afflicted with a singular trouble, which seems to have been due to an extraordinary excitability of the retina. It was the appearance of images which, by their persistence, marred the vision of real objects and interfered with thought. When a word was said to me, the image of the object which it designated would appear vividly before my eyes, and many times it was impossible for me to tell whether the object I saw was real or not. This caused me great discomfort and anxiety, and I tried hard to free myself of the spell. But for a long time I tried in vain, and it was not, as I still clearly recollect, until I was about twelve years old that I succeeded for the first time, by an effort of the will, in banishing an image which presented itself. My happiness will never be as complete as it was then, but, unfortunately (as I thought at that time), the old trouble returned, and with it my anxiety. Here it was that the observations to which I refer began. I noted, namely, that whenever the image of an object appeared before my eyes I had seen something that reminded me of it. In the first instances I thought this to be purely accidental, but soon I convinced myself that it was not so. A visual impression, consciously or unconsciously received, invariably preceded the appearance of the image. Gradually the desire arose in me to find out, every time, what caused the images to appear, and the satisfaction of this desire soon became a necessity. The next observation I made was that, just as these images followed as a result of something I had seen, so also the thoughts which I conceived were suggested in like manner. Again, I experienced the same desire to locate the image which caused the thought, and this search for the original visual impression soon grew to be a second nature. My mind became automatic, as it were, and in the course of years of continued, almost unconscious performance, I acquired the ability of locating every time and, as a rule, instantly the visual impression which started the thought. Nor is this all. It was not long before I was aware that also all my movements were prompted in the same way, and so, searching, observing, and verifying continuously, year by year, I have, by every thought and every act of mine, demonstrated, and do so daily, to my absolute satisfaction, that I am an automaton endowed with power of movement, which merely responds to external stimuli beating upon my sense organs, and thinks and acts and moves accordingly. I remember only one or two cases in all my life in which I was unable to locate the first impression which prompted a movement or a thought, or even a dream.

Fig. 2. — The First Practical Telautomaton.

Note to Fig. 2. — A machine having all the bodily or translatory movements and the operations of the interior mechanism controlled from a distance without wires. The crewless boat shown in the photograph contains its own motive power, propelling and steering machinery, and numerous other accessories, all of which are controlled by transmitting from a distance, without wires, electrical oscillations to a circuit carried by the boat and adjusted to respond only to these oscillations.

With these experiences it was only natural that, long ago, I conceived the idea of constructing an automaton which would mechanically represent me, and which would respond, as I do myself, but, of course, in a much more primitive manner, to external influences. Such an automaton evidently had to have motive power, organs for locomotion, directive organs, and one or more sensitive organs so adapted as to be excited by external stimuli. This machine would, I reasoned, perform its movements in the manner of a living being, for it would have all the chief mechanical characteristics or elements of the same. There was still the capacity for growth, propagation, and, above all, the mind which would be wanting to make the model complete. But growth was not necessary in this case, since a machine could be manufactured full-grown, so to speak. As to the capacity for propagation, it could likewise be left out of consideration, for in the mechanical model it merely signified a process of manufacture. Whether the automaton be of flesh and bone, or of wood and steel, it mattered little, provided it could perform all the duties required of it like an intelligent being. To do so, it had to have an element corresponding to the mind, which would effect the control of all its movements and operations, and cause it to act, in any unforeseen case that might present itself, with knowledge, reason, judgment, and experience. But this element I could easily embody in it by conveying to it my own intelligence, my own understanding. So this invention was evolved, and so a new art came into existence, for which the name "telautomatics" has been suggested, which means the art of controlling the movements and operations of distant automatons.

Fig. 3. — Experiment to illustrate the supplying of electrical energy through a single wire without return.

Note to Fig. 3. — An ordinary incandescent lamp, connected with one or both of its terminals to the wire forming the upper free end of the coil shown in the photograph, is lighted by electrical vibrations conveyed to it through the coil from an electrical oscillator, which is worked only to one fifth of one per cent. of its full capacity.

This principle evidently was applicable to any kind of machine that moves on land or in the water or in the air. In applying it practically for the first time, I selected a boat (see Fig. 2). A storage battery placed within it furnished the motive power. The propeller, driven by a motor, represented the locomotive organs. The rudder, controlled by another motor likewise driven by the battery, took the place of the directive organs. As to the sensitive organ, obviously the first thought was to utilize a device responsive to rays of light, like a selenium cell, to represent the human eye. But upon closer inquiry I found that, owing to experimental and other difficulties, no thoroughly satisfactory control of the automaton could be effected by light, radiant heat, Hertzian radiations, or by rays in general, that is, disturbances which pass in straight lines through space. One of the reasons was that any obstacle coming between the operator and the distant automaton would place it beyond his control. Another reason was that the sensitive device representing the eye would have to be in a definite position with respect to the distant controlling apparatus, and this necessity would impose great limitations in the control. Still another and very important reason was that, in using rays, it would be difficult, if not impossible, to give to the automaton individual features or characteristics distinguishing it from other machines of this kind. Evidently the automaton should respond only to an individual call, as a person responds to a name. Such considerations led me to conclude that the sensitive device of the machine should correspond to the ear rather than to the eye of a human being, for in this case its actions could be controlled irrespective of intervening obstacles, regardless of its position relative to the distant controlling apparatus, and, last, but not least, it would remain deaf and unresponsive, like a faithful servant, to all calls but that of its master. These requirements made it imperative to use, in the control of the automaton, instead of light or other rays, waves or disturbances which propagate in all directions through space, like sound, or which follow a path of least resistance, however curved. I attained the result aimed at by means of an electric circuit placed within the boat, and adjusted, or "tuned," exactly to electrical vibrations of the proper kind transmitted to it from a distant "electrical oscillator." This circuit, in responding, however feebly, to the transmitted vibrations, affected magnets and other contrivances, through the medium of which were controlled the

161

movements of the propeller and rudder, and also the operations of numerous other appliances.

Fig. 4. — Experiment to illustrate the transmission of electrical energy through the earth without wire.

By the simple means described the knowledge, experience, judgment — the mind, so to speak — of the distant operator were embodied in that machine, which was thus enabled to move and to perform all its operations with reason and intelligence. It behaved just like a blindfolded person obeying directions received through the ear.

Note to Fig. 4. — The coil shown in the photograph has its lower end or terminal connected to the ground, and is exactly attuned to the vibrations of a distant electrical oscillator. The lamp lighted is an independent wire loop, energized by induction from the coil excited by the electrical vibrations transmitted to it through the ground from the oscillator, which is worked only to five per cent. of its full capacity.

The automatons so far constructed had "borrowed minds," so to speak, as each merely formed part of the distant operator who conveyed to it his intelligent orders; but this art is only in the beginning. I purpose to show that, however impossible it may now seem, an automaton may be contrived which will have its "own mind," and by this I mean that it will be able, independent of any operator, left entirely to itself, to perform, in response to external influences affecting its sensitive organs, a great variety of acts and operations as if it had intelligence. It will be able to follow a course laid out or to obey orders given far in advance; it will be capable of distinguishing between what it ought and what it ought not to do, and of making experiences or, otherwise stated, of recording impressions which will definitely affect its subsequent actions. In fact, I have already conceived such a plan.

Fig. 5. — Photographic view of coils responding to electrical oscillations.

Note to Fig. 5. — The picture shows a number of coils, differently attuned and responding to the vibrations transmitted to them through the earth from an electrical oscillator. The large coil on the right, discharging strongly, is tuned to the fundamental vibration, which is fifty thousand per second; the two larger vertical coils to twice that number; the smaller white wire coil to four times that number, and the remaining small coils to higher tones. The vibrations produced by the oscillator were so intense that they affected perceptibly a small coil tuned to the twenty-sixth higher tone.

Although I evolved this invention many years ago and explained it to my visitors very frequently in my laboratory demonstrations, it was not until much later, long after I had perfected it, that it became known, when, naturally enough, it gave rise to much discussion and to sensational reports. But the true

165

significance of this new art was not grasped by the majority, nor was the great force of the underlying principle recognized. As nearly as I could judge from the numerous comments which appeared, the results I had obtained were considered as entirely impossible. Even the few who were disposed to admit the practicability of the invention saw in it merely an automobile torpedo, which was to be used for the purpose of blowing up battle-ships, with doubtful success. The general impression was that I contemplated simply the steering of such a vessel by means of Hertzian or other rays. There are torpedoes steered electrically by wires, and there are means of communicating without wires, and the above was, of course, an obvious inference. Had I accomplished nothing more than this, I should have made a small advance indeed. But the art I have evolved does not contemplate merely the change of direction of a moving vessel; it affords means of absolutely controlling, in every respect, all the innumerable translatory movements, as well as the operations of all the internal organs, no matter how many, of an individualized automaton. Criticisms to the effect that the control of the automaton could be interfered with were made by people who do not even dream of the wonderful results which can be accomplished by use of electrical vibrations. The world moves slowly, and new truths are difficult to see. Certainly, by the use of this principle, an arm for attack as well as defense may be provided, of a destructiveness all the greater as the principle is applicable to submarine and aërial vessels. There is virtually no restriction as to the amount of explosive it can carry, or as to the distance at which it can strike, and failure is almost impossible. But the force of this new principle does not wholly reside in its destructiveness. Its advent introduces into warfare an element which never existed before — a fighting-machine without men as a means of attack and defense. The continuous development in this direction must ultimately make war a mere contest of machines without men and without loss of life — a condition which would have been impossible without this new departure, and which, in my opinion, must be reached as preliminary to permanent peace. The future will either bear out or disprove these views. My ideas on this subject have been put forth with deep conviction, but in a humble spirit.

Fig. 6. — Photographic view of the essential parts of the electrical oscillator used in the experiments described.

The establishment of permanent peaceful relations between nations would most effectively reduce the force retarding the human mass, and would be the best solution of this great human problem. But will the dream of universal peace ever be realized? Let us hope that it will. When all darkness shall be dissipated by the light of science, when all nations shall be merged into one, and patriotism shall be identical with religion, when there shall be one language, one country, one end, then the dream will have become reality.

THE THIRD PROBLEM: HOW TO INCREASE THE FORCE ACCELERATING THE HUMAN MASS — THE HARNESSING OF THE SUN'S ENERGY.

Of the three possible solutions of the main problem of increasing human energy, this is by far the most important to consider, not only because of its intrinsic significance, but also because of its intimate bearing on all the many elements and conditions which determine the movement of humanity. In order to proceed systematically, it would be necessary for me to dwell on all those considerations which have guided me from the outset in my efforts to arrive at a solution, and which have led me, step by step, to the results I shall now describe. As a preliminary study of the problem an analytical investigation, such as I have made, of the chief forces which determine the onward movement, would be of advantage, particularly in conveying an idea of that hypothetical "velocity" which, as explained in the beginning, is a measure of human energy; but to deal with this specifically here, as I would desire, would lead me far beyond the scope of the present subject. Suffice it to state that the resultant of all these forces is always in the direction of reason, which, therefore, determines, at any time, the direction of human movement. This is to say that every effort which is scientifically applied, rational, useful, or practical, must be in the direction in which the mass is moving. The practical, rational man, the observer, the man of business, he who reasons, calculates, or determines in advance, carefully applies his effort so that when coming into effect it will be in the direction of the movement, making it thus most efficient, and in this knowledge and ability lies the secret of his success. Every new fact discovered, every new experience or new element added to our knowledge and entering into the domain of reason, affects the same and, therefore, changes the direction of movement, which, however, must always take place along the resultant of all those efforts which, at that time, we designate as reasonable, that is, self-preserving, useful, profitable, or practical. These efforts concern our daily life, our necessities and comforts, our work and business, and it is these which drive man onward.

Fig. 7. — Experiment to illustrate an inductive effect of an electrical oscillator of great power.

Note to Fig. 7. — The photograph shows three ordinary incandescent lamps lighted to full candle-power by currents induced in a local loop consisting of a single wire forming a square of fifty feet each side, which includes the lamps, and which is at a distance of one hundred feet from the primary circuit

169

energized by the oscillator. The loop likewise includes an electrical condenser, and is exactly attuned to the vibrations of the oscillator, which is worked at less than five per cent. of its total capacity.

Fig. 8. — Experiment to illustrate the capacity of the oscillator for producing electrical explosions of great power.

Note to Fig. 8. — The coil, partly shown in the photograph, creates an alternative movement of electricity from the earth into a large reservoir and back at a rate of one hundred thousand alternations per second. The adjustments are such that the reservoir is filled full and bursts at each alternation just at the moment when the electrical pressure reaches the maximum. The discharge escapes with a deafening noise, striking an unconnected coil twenty-two feet away, and creating such a commotion of electricity in the earth that sparks an inch long can be drawn from a water main at a distance of three hundred feet from the laboratory.

Fig. 9. — Experiment to illustrate the capacity of the oscillator for creating a great electrical movement.

Note to Fig. 9. — The ball shown in the photograph, covered with a polished metallic coating of twenty square feet of surface, represents a large reservoir of electricity, and the inverted tin pan underneath, with a sharp rim, a big opening through which the electricity can escape before filling the reservoir. The quantity of electricity set in movement is so great that, although most of it escapes through the rim of the pan or opening provided, the ball or reservoir is nevertheless alternately emptied and filled to over-flowing (as is evident from the discharge escaping on the top of the ball) one hundred and fifty thousand times per second.

But looking at all this busy world about us, on all this complex mass as it daily throbs and moves, what is it but an immense clock-work driven by a spring? In the morning, when we rise, we cannot fail to note that all the objects about us are manufactured by machinery: the water we use is lifted by steam-power; the trains bring our breakfast from distant localities; the elevators in our dwelling and our office building, the cars that carry us there, are all driven by power; in all our daily errands, and in our very life-pursuit, we depend upon it; all the objects we see tell us of it; and when we return to our machine-made dwelling at night, lest we should forget it, all the material comforts of our home, our cheering stove and lamp, remind us of how much we depend on power. And when there is an accidental stoppage of the machinery, when the city is snow-bound, or the life-sustaining movement otherwise temporarily arrested, we are affrighted to realize how impossible it would be for us to live the life we live without motive power. Motive power means work. To increase the force accelerating human movement means, therefore, to perform more work.

Fig. 10. — Photographic view of an experiment to illustrate an effect of an electrical oscillator delivering energy at a rate of seventy-five thousand horse-power.

Note to Fig. 10. — The discharge, creating a strong draft owing to the heating of the air, is carried upward through the open roof of the building. The greatest width across is nearly seventy feet. The pressure is over twelve million volts, and the current alternates one hundred and thirty thousand times per second.

So we find that the three possible solutions of the great problem of increasing human energy are answered by the three words: *food*, *peace*, *work*. Many a year I have thought and pondered, lost myself in speculations and theories, considering man as a mass moved by a force, viewing his inexplicable movement in the light of a mechanical one, and applying the simple principles of mechanics to the analysis of the same until I arrived at these solutions, only to realize that they were taught to me in my early childhood. These three words sound the key-notes of the Christian religion. Their scientific meaning and purpose now clear to me: food to increase the mass, peace to diminish the retarding force, and work to increase the force accelerating human movement. These are the only three solutions which are possible of that great problem, and all of them have one object, one end, namely, to increase human energy. When we recognize this, we cannot help wondering how profoundly wise and scientific and how immensely practical the Christian religion is, and in what a marked contrast it stands in this respect to other religions. It is unmistakably the result of practical experiment and scientific observation which have extended through the ages, while other religions seem to be the outcome of merely abstract reasoning. Work, untiring effort, useful and accumulative, with periods of rest and recuperation aiming at higher efficiency, is its chief and ever-recurring command. Thus we are inspired both by Christianity and Science to do our utmost toward increasing the performance of mankind. This most important of human problems I shall now specifically consider.

THE SOURCE OF HUMAN ENERGY — THE THREE WAYS OF DRAWING ENERGY FROM THE SUN.

First let us ask: Whence comes all the motive power? What is the spring that drives all? We see the ocean rise and fall, the rivers flow, the wind, rain, hail, and snow beat on our windows, the trains and steamers come and go; we here the rattling noise of carriages, the voices from the street; we feel, smell, and taste; and we think of all this. And all this movement, from the surging of the mighty ocean to that subtle movement concerned in our thought, has but one common cause. All this energy emanates from one single center, one single source — the sun. The sun is the spring that drives all. The sun maintains all human life and supplies all human energy. Another answer we have now found to the above great question: To increase the force accelerating human movement means to turn to the uses of man more of the sun's energy. We honor and revere those great men of bygone times whose names are linked with immortal achievements, who have proved themselves benefactors of humanity — the religious reformer with his wise maxims of life, the philosopher with his deep truths, the mathematician with his formulæ, the physicist with his laws, the discover with his principles and secrets wrested from nature, the artist with his forms of the beautiful; but who honors him, the greatest of all, — who can tell the name of him, — who first turned to use the sun's energy to save the effort of a weak fellow-creature? That was man's first act of scientific philanthropy, and its consequences have been incalculable.

From the very beginning three ways of drawing energy from the sun were open to man. The savage, when he warmed his frozen limbs at a fire kindled in some way, availed himself of the energy of the sun stored in the burning material. When he carried a

175

bundle of branches to his cave and burned them there, he made use of the sun's stored energy transported from one to another locality. When he set sail to his canoe, he utilized the energy of the sun applied to the atmosphere or the ambient medium. There can be no doubt that the first is the oldest way. A fire, found accidentally, taught the savage to appreciate its beneficial heat. He then very likely conceived of the idea of carrying the glowing members to his abode. Finally he learned to use the force of a swift current of water or air. It is characteristic of modern development that progress has been effected in the same order. The utilization of the energy stored in wood or coal, or, generally speaking, fuel, led to the steam-engine. Next a great stride in advance was made in energy-transportation by the use of electricity, which permitted the transfer of energy from one locality to another without transporting the material. But as to the utilization of the energy of the ambient medium, no radical step forward has as yet been made known.

The ultimate results of development in these three directions are: first, the burning of coal by a cold process in a battery; second, the efficient utilization of the energy of the ambient medium; and, third, the transmission without wires of electrical energy to any distance. In whatever way these results may be arrived at, their practical application will necessarily involve an extensive use of iron, and this invaluable metal will undoubtedly be an essential element in the further development along these three lines. If we succeed in burning coal by a cold process and thus obtain electrical energy in an efficient and inexpensive manner, we shall require in many practical uses of this energy electric motors — that is, iron. If we are successful in deriving energy from the ambient medium, we shall need, both in the obtainment and utilization of the energy, machinery

— again, iron. If we realize the transmission of electrical energy without wires on an industrial scale, we shall be compelled to use extensively electric generators — once more, iron. Whatever we may do, iron will probably be the chief means of accomplishment in the near future, possibly more so than in the past. How long its reign will last is difficult to tell, for even now aluminium is looming up as a threatening competitor. But for the time being, next to providing new resources of energy, it is of the greatest importance to making improvements in the manufacture and utilization of iron. Great advances are possible in these latter directions, which, if brought about, would enormously increase the useful performance of mankind.

GREAT POSSIBILITIES OFFERED BY IRON FOR INCREASING HUMAN PERFORMANCE— ENORMOUS WASTE IN IRON MANUFACTURE.

Iron is by far the most important factor in modern progress. It contributes more than any other industrial product to the force accelerating human movement. So general is the use of this metal, and so intimately is it connected with all that concerns our life, that it has become as indispensable to us as the very air we breathe. Its name is synonymous with usefulness. But, however great the influence of iron may be on the present human development, it does not add to the force urging man onward nearly as much as it might. First of all, its manufacture as now carried on is connected with an appalling waste of fuel — that is, waste of energy. Then, again, only a part of all the iron produced is applied for useful purposes. A good part of it goes to create frictional resistances, while still another large part is the means of developing negative forces greatly retarding human movement. Thus the negative force of war is almost wholly represented in iron. It is impossible to estimate with any degree of accuracy the magnitude of this greatest of all retarding forces, but it is certainly very considerable. If the present positive impelling force due to all useful applications of iron be represented by ten, for instance, I should not think it exaggeration to estimate the negative force of war, with due consideration of all its retarding influences and results, at, say, six. On the basis of this estimate the effective impelling force of iron in the positive direction would be measured by the difference of these two numbers, which is four. But if, through the establishment of universal peace, the manufacture of war machinery should cease, and all struggle for supremacy between nations should be turned into healthful, ever active and productive commercial competition, then the positive impelling force due to iron would be measured by the sum of those two, numbers, which is sixteen — that is, this force would have four times its present value. This example is, of course, merely intended to give an idea of the immense increase in the useful performance of mankind which would result from a radical reform of the iron industries supplying the implements of warfare.

A similar inestimable advantage in the saving of energy available to man would be secured by obviating the great waste of coal which is inseparably connected with the present methods of manufacturing iron. In some countries, as in Great Britain, the hurtful effects of this squandering of fuel are beginning to be felt. The price of coal is constantly rising, and the poor are made to suffer more and more. Though we are still far from the dreaded "exhaustion of the coal-fields," philanthropy commands us to invent novel methods of manufacturing iron, which will not involve such barbarous waste of this valuable material from which we derive at present most of our energy. It is our duty to coming generations to leave this store of energy intact for them, or at least not to touch it until we shall have perfected processes for burning coal more efficiently. Those who are coming after us will need fuel more than we do. We should be able to manufacture the iron we require by using the sun's energy, without wasting any coal at all. As an effort to this end the idea of smelting iron ores by electric currents obtained from the energy of falling water has naturally suggested itself to many. I have myself spent much time in endeavoring to evolve such a practical process, which would enable iron to be manufactured at small cost. After a prolonged investigation of the subject, finding that it was unprofitable to use the currents generated directly for smelting the ore, I devised a method which is far more economical.

ECONOMICAL PRODUCTION OF IRON BY A NEW PROCESS.

The industrial project, as I worked it out six years ago, contemplated the employment of the electric currents derived from the energy of a waterfall, not directly for smelting the ore, but for decomposing water, as a preliminary step. To lessen the cost of the plant, I proposed to generate the currents in exceptionally cheap and simple dynamos, which I designed for this sole purpose. The hydrogen liberated in the electrolytic decomposition was to be burned or recombined with oxygen, not with that from which it was separated, but with that of the atmosphere. Thus very nearly the total electrical energy used up in the decomposition of the water would be recovered in the form of heat resulting from the recombination of the hydrogen. This heat was to be applied to the smelting of ore. The oxygen gained as a by-product of the decomposition of the water I intended to use for certain other industrial purposes, which would probably yield good financial returns, inasmuch as this is the cheapest way of obtaining this gas in large quantities. In any event, it could be employed to burn all kinds of refuse, cheap hydrocarbon, or coal of the most inferior quality which could not be burned in air or be otherwise utilized to advantage, and thus again a considerable amount of heat would be made available for the smelting of the ore. To increase the economy of the process I contemplated, furthermore, using an arrangement such that the hot metal and the products of combustion, coming out of the furnace, would give up their heat upon the cold ore going into the furnace, so that comparatively little of the heat-energy would be lost in the smelting. I calculated that probably forty thousand pounds of iron could be produced per horse-power per annum by this method. Liberal allowances were made for those losses which are unavoidable, the above quantity being about half of that theoretically obtainable. Relying on this estimate and on practical data with reference to a certain kind of sand ore existing in abundance in the region of the Great Lakes, including cost of transportation and labor, I found that in some localities iron could be manufactured in this manner cheaper than by any of the adopted methods. This result would be attained all the more surely if the oxygen obtained from the water, instead of being used for smelting of ore, as assumed, should be more profitably employed. Any new demand for this gas would secure a higher revenue from the plant, thus cheapening the iron. This project was advanced merely in the interest of industry. Some day, I hope, a beautiful industrial butterfly will come out of the dusty and shriveled chrysalis.

The production of iron from sand ores by a process of magnetic separation is highly commendable in principle, since it involves no waste of coal; but the usefulness of this method is largely reduced by the necessity of melting the iron afterward. As to the crushing of iron ore, I would consider it rational only if done by water-power, or by energy otherwise obtained without consumption of fuel. An electrolytic cold process, which would make it possible to extract iron cheaply, and also to mold it into the required forms without any fuel consumption, would, in my opinion, be a very great advance in iron manufacture. In common with some other metals, iron has so far resisted electrolytic treatment, but there can be no doubt that such a cold process will ultimately replace in metallurgy the present crude method of casting, and thus obviating the enormous waste of fuel necessitated by the repeated heating of metal in the foundries.

Up to a few decades ago the usefulness of iron was based almost wholly on its remarkable mechanical properties, but since the advent of the commercial dynamo and electric motor its value to mankind has been greatly increased by its unique magnetic qualities. As regards the latter, iron has been greatly improved of late. The signal progress began about thirteen years ago, when I discovered that in using soft Bessemer steel instead of wrought iron, as then customary, in an alternating motor, the performance of the machine was doubled. I brought this fact to the attention of Mr. Albert Schmid, to whose untiring efforts and ability is largely due the supremacy of American electrical machinery, and who was then superintendent of an industrial corporation engaged in this field. Following my suggestion, he constructed transformers of steel, and they showed the same marked improvement. The investigation was then systematically continued under Mr. Schmid's guidance, the impurities being gradually eliminated from the "steel" (which was only such in name, for in reality it was pure soft iron), and soon a product resulted which admitted of little further improvement.

THE COMING OF AGE OF ALUMINIUM — DOOM OF THE COPPER INDUSTRY — THE GREAT CIVILIZING POTENCY OF THE NEW METAL.

With the advances made in iron of late years we have arrived virtually at the limits of improvement. We cannot hope to increase very materially its tensile strength, elasticity, hardness, or malleability, nor can we expect to make it much better as regards its magnetic qualities. More recently a notable gain was secured by the mixture of a small percentage of nickel with the iron, but there is not much room for further advance in this direction. New discoveries may be expected, but they cannot greatly add to the valuable properties of the metal, though they may considerably reduce the cost of manufacture. The immediate future of iron is assured by its cheapness and its unrivaled mechanical and magnetic qualities. These are such that no other product can compete with it now. But there can be no doubt that, at a time not very distant, iron, in many of its now uncontested domains, will have to pass the scepter to another: the coming age will be the age of aluminium. It is only seventy years since this wonderful metal was discovered by Woehler, and the aluminium industry, scarcely forty years old, commands already the attention of the entire world. Such rapid growth has not been recorded in the history of civilization before. Not long ago aluminium was sold at the fanciful price of thirty or forty dollars per pound; to-day it can be had in any desired amount for as many cents. What is more, the time is not far off when this price, too, will be considered fanciful, for great improvements are possible in the methods of its manufacture. Most of the metal is now produced in the electric furnace by a process combining fusion and electrolysis, which offers a number of advantageous features, but involves naturally a great waste of the electrical energy of the current. My estimates show that the price of aluminium could be considerably reduced by adopting in its manufacture a method similar to that proposed by me for the production of iron. A pound of aluminium requires for fusion only about seventy per cent. of the heat needed for melting a pound of iron, and inasmuch as its weight is only about one third of that of the latter, a volume of aluminium four times that of iron could be obtained from a given amount of heat-energy. But a cold electrolytic process of manufacture is the ideal solution, and on this I have placed my hope.

The absolutely unavoidable consequence of the advance of the aluminium industry will be the annihilation of the copper industry. They cannot exist and prosper together, and the latter is doomed beyond any hope of recovery. Even now it is cheaper to convey an electric current through aluminium wires than through copper wires; aluminium castings cost less, and in many domestic and other uses copper has no chance of successfully competing. A further material reduction of the price of aluminium cannot but be fatal to copper. But the progress of the former will not go on unchecked, for, as it ever happens in such cases, the larger industry will absorb the smaller one: the giant copper interests will control the pygmy aluminium interests, and the slow-pacing copper will reduce the lively gait of aluminium. This will only delay, not avoid, the impending catastrophe.

Aluminium, however, will not stop at downing copper. Before many years have passed it will be engaged in a fierce struggle with iron, and in the latter it will find an adversary not easy to conquer. The issue of the contest will largely depend on whether iron shall be indispensable in electric machinery. This the future alone can decide. The magnetism as exhibited in iron is an isolated phenomenon in nature. What it is that makes this metal behave so radically different from all other materials in this respect has not yet been ascertained, though many theories have been suggested. As regards magnetism, the molecules of the various bodies behave like hollow beams partly filled with a heavy fluid and balanced in the middle in the manner of a see-saw. Evidently some disturbing influence exists in nature which causes each molecule, like such a beam, to tilt either one or the other way. If the molecules are tilted one way, the body is magnetic; if they are tilted the other way, the body is non-magnetic; but both positions are stable, as they would be in the case of the hollow beam, owing to the rushing of the fluid to the lower end. Now, the wonderful thing is that the molecules of all known bodies went one way, while those of iron went the other way. This metal, it would seem, has an origin entirely different from that of the rest of the globe. It is highly improbable that we shall discover some other and cheaper material which will equal or surpass iron in magnetic qualities.

Unless we should make a radical departure in the character of the electric currents employed, iron will be indispensable. Yet the advantages it offers are only apparent. So long as we use feeble magnetic forces it is by far superior to any other material; but if we find ways of producing great magnetic forces, than better results will be obtainable without it. In fact, I have already produced electric transformers in which no iron is employed, and which are capable of performing ten times as much work per pound of weight as those with iron. This result is attained by using electric currents of a very high rate of vibration, produced in novel ways, instead of the ordinary currents now employed in the industries. I have also succeeded in operating electric motors without iron by such rapidly vibrating currents, but the results, so far, have been inferior to those obtained with ordinary motors constructed of iron, although theoretically the former should be capable of performing incomparably more work per unit of weight than the latter. But the seemingly insuperable difficulties which are now in the way may be overcome in the end, and then iron will be done away with, and all electric machinery will be manufactured of aluminium, in all probability, at prices ridiculously low. This would be a severe, if not fatal, blow to iron. In many other branches of industry, as ship-building, or wherever lightness of structure is required, the progress of the new metal will be much quicker. For such uses it is eminently suitable, and is sure to supersede iron sooner or later. It is highly probable that in the course of time we shall be able to give it many of those qualities which make iron so valuable.

While it is impossible to tell when this industrial revolution will be consummated, there can be no doubt that the future belongs to aluminium, and that in times to come it will be the chief means of increasing human performance. It has in this respect capacities greater by far than those of any other metal. I should estimate its civilizing potency at fully one hundred times that of iron. This estimate, though it may astonish, is not at all exaggerated. First of all, we must remember that there is thirty times as much aluminium as iron in bulk, available for the uses of man. This in itself offers great possibilities. Then, again, the new metal is much more easily workable, which adds to its value. In many of its properties it partakes of the character of a precious metal, which gives it additional worth. Its electric conductivity, which, for a given weight, is greater than that of any other metal, would be alone sufficient to make it one of the most important factors in future human progress. Its extreme lightness makes it far more easy to transport the objects manufactured. By virtue of this property it will revolutionize naval construction, and in facilitating transport and travel it will add enormously to the useful performance of mankind. But its greatest civilizing property will be, I believe, in aërial travel, which is sure to be brought about by means of it. Telegraphic instruments will slowly enlighten the barbarian. Electric motors and lamps will do it more quickly, but quicker than anything else the flying-machine will do it. By rendering travel ideally easy it will be the best means for unifying the heterogeneous elements of humanity. As the first step toward this realization we should produce a lighter storage-battery or get more energy from coal.

EFFORTS TOWARD OBTAINING MORE ENERGY FROM COAL — THE ELECTRIC TRANSMISSION — THE GAS-ENGINE — THE COLD-COAL BATTERY.

I remember that at one time I considered the production of electricity by burning coal in a battery as the greatest achievement toward the advancing civilization, and I am surprised to find how much the continuous study of these subjects has modified my views. It now seems to me that to burn coal, however efficiently, in a battery would be a mere makeshift, a phase in the evolution toward something much more perfect. After all, in generating electricity in this manner, we should be destroying material, and this would be a barbarous process. We ought to be able to obtain the energy we need without consumption of material. But I am far from underrating the value of such an efficient method of burning fuel. At the present time most motive power comes from coal, and, either directly or by its products, it adds vastly to human energy. Unfortunately, in all the processes now adopted, the larger portion of the energy of the coal is uselessly dissipated. The best steam-engines utilize only a small part of the total energy. Even in gas-engines, in which, particularly of late, better results are obtainable, there is still a barbarous waste going on. In our electric-lighting systems we scarcely utilize one third of one per cent., and in lighting by gas a much smaller fraction, of the total energy of the coal. Considering the various uses of coal throughout the world, we certainly do not utilize more than two per cent. of its energy theoretically available. The man who should stop this senseless waste would be a great benefactor of humanity, though the solution he would offer could not be a permanent one, since it would ultimately lead to the exhaustion of the store of material. Efforts toward obtaining more energy from coal are now being made chiefly in two directions — by generating electricity and by producing gas for motive-power purposes. In both of these lines notable success has already been achieved.

The advent of the alternating-current system of electric power-transmission marks an epoch in the economy of energy available to man from coal. Evidently all electrical energy obtained from a waterfall, saving so much fuel, is a net gain to mankind, which is all the more effective as it is secured with little expenditure of human effort, and as this most perfect of all known methods of deriving energy from the sun contributes in many ways to the advancement of civilization. But electricity enables us also to get from coal much more energy than was practicable in the old ways. Instead of transporting the coal to distant places of consumption, we burn it near the mine, develop electricity in the dynamos, and transmit the current to remote localities, thus effecting a considerable saving. Instead of driving the machinery in a factory in the old wasteful way of belts and shafting, we generate electricity by steam-power and operate electric motors. In this manner it is not uncommon to obtain two or three times as much effective motive power from the fuel, besides securing many other important advantages. It is in this field as much as in the transmission of energy to great distances that the alternating system, with its ideally simple machinery, is bringing about an industrial revolution. But in many lines this progress has not been yet fully felt. For example, steamers and trains are still being propelled by the direct application of steam-power to shafts or axles. A much greater percentage of the heat-energy of the fuel could be transformed into motive energy by using, in place of the adopted marine engines and locomotives, dynamos driven by specially designed high-pressure steam- or gas-engines and by utilizing the electricity generated for the propulsion. A gain of fifty to one hundred per cent. in the effective energy derived from the coal could be secured in this manner. It is difficult to understand why a fact so plain and obvious is not receiving more attention from engineers. In ocean steamers such an improvement would be particularly desirable, as it would do away with noise and increase materially the speed and the carrying capacity of the liners.

Still more energy is now being obtained from coal by the latest improved gas-engine, the economy of which is, on the average, probably twice that of the best steam-engine. The introduction of the gas-engine is very much facilitated by the importance of the gas industry. With the increasing use of the electric light more and more of the gas is utilized for heating and motive-power purposes. In many instances gas is manufactured close to the coal-mine and conveyed to distant places of consumption, a considerable saving both in the cost of transportation and in utilization of the energy of the fuel being thus effected. In the present state of the mechanical and electrical arts the most rational way of deriving energy from coal is evidently to manufacture gas close to the coal store, and to utilize it, either on the spot or elsewhere, to generate electricity for industrial uses in dynamos driven by gas engines. The commercial success of such a plant is largely dependent upon the production of gas-engines of great nominal horse-power, which, judging from the keen activity in this field, will soon be forthcoming. Instead of consuming coal directly, as usual, gas should be manufactured from it and burned to economize energy.

But all such improvements cannot be more than passing phases in the evolution toward something far more perfect, for ultimately we must succeed in obtaining electricity from coal in a more direct way, involving no great loss of heat-energy. Whether coal can be oxidized by a cold process is still a question. Its combination with oxygen always evolves heat, and whether the energy of the combination of the carbon with another element can be turned directly into electrical energy has not yet been determined. Under certain conditions nitric acid will burn the carbon, generating an electric current, but the solution does not remain cold. Other means of oxidizing coal have been proposed, but they have offered no promise of leading to an efficient process. My own lack of success has been complete, though perhaps not quite so complete as that of some who have "perfected" the cold-coal battery. This problem is essentially one for the chemist to solve. It is not for the physicist, who determines all his results in advance, so that, when the experiment is tried, it cannot fail. Chemistry, though a positive science, does not yet admit of a solution by such positive methods as those which are available in the treatment of many physical problems. The result, if possible, will be arrived at through patient trying rather than through deduction or calculation. The time will soon come, however, when the chemist will be able to follow a course clearly mapped out beforehand, and when the process of his arriving at a desired result will be purely constructive. The cold-coal battery would give a great impetus to electrical development; it would lead very shortly to a practical flying-machine, and would enormously enhance the introduction of the automobile. But these and many other problems will be better solved, and in a more scientific manner, by a light-storage battery.

ENERGY FROM THE MEDIUM—THE WINDMILL AND THE SOLAR ENGINE,—MOTIVE POWER FROM TERRESTRIAL HEAT—ELECTRICITY FROM NATURAL SOURCES.

Besides fuel, there is abundant material from which we might eventually derive power. An immense amount of energy is locked up in limestone, for instance, and machines can be driven by liberating the carbonic acid through sulphuric acid or otherwise. I once constructed such an engine, and it operated satisfactorily.

But, whatever our resources of primary energy may be in the future, we must, to be rational, obtain it without consumption of any material. Long ago I came to this conclusion, and to arrive at this result only two ways, as before indicated, appeared possible — either to turn to use the energy of the sun stored in the ambient medium, or to transmit, through the medium, the sun's energy to distant places from some locality where it was obtainable without consumption of material. At that time I at once rejected the latter method as entirely impracticable, and turned to examine the possibilities of the former.

It is difficult to believe, but it is, nevertheless, a fact, that since time immemorial man has had at his disposal a fairly good machine which has enabled him to utilize the energy of the ambient medium. This machine is the windmill. Contrary to popular belief, the power obtainable from wind is very considerable. Many a deluded inventor has spent years of his life in endeavoring to "harness the tides," and some have even proposed to compress air by tide- or wave-power for supplying energy, never understanding the signs of the old windmill on the hill, as it sorrowfully waved its arms about and bade them stop. The fact is that a wave- or tide-motor would have, as a rule, but a small chance of competing commercially with the windmill, which is by far the better machine, allowing a much greater amount of energy to be obtained in a simpler way. Wind-power has been, in old times, of inestimable value to man, if for nothing else but for enabling him to cross the seas, and it is even now a very important factor in travel and transportation. But there are great limitations in this ideally simple method of utilizing the sun's energy. The machines are large for a given output, and the power is intermittent, thus necessitating the storage of energy and increasing the cost of the plant.

A far better way, however, to obtain power would be to avail ourselves of the sun's rays, which beat the earth incessantly and supply energy at a maximum rate of over four million horsepower per square mile. Although the average energy received per square mile in any locality during the year is only a small fraction of that amount, yet an inexhaustible source of power would be opened up by the discovery of some efficient method of utilizing the energy of the rays. The only rational way known to me at the time when I began the study of this subject was to employ some kind of heat- or thermodynamic engine, driven by a volatile fluid evaporate in a boiler by the heat of the rays. But closer investigation of this method, and calculation, showed that, notwithstanding the apparently vast amount of energy received from the sun's rays, only a small fraction of that energy could be actually utilized in this manner. Furthermore, the energy supplied through the sun's radiations is periodical, and the same limitations as in the use of the windmill I found to exist here also. After a long study of this mode of obtaining motive power from the sun, taking into account the necessarily large bulk of the boiler, the low efficiency of the heat-engine, the additional cost of storing the energy, and other drawbacks, I came to the conclusion that the "solar engine," a few instances excepted, could not be industrially exploited with success.

Another way of getting motive power from the medium without consuming any material would be to utilize the heat contained in the earth, the water, or the air for driving an engine. It is a well-known fact that the interior portions of the globe are very hot, the temperature rising, as observations show, with the approach to the center at the rate of approximately 1° C. for every hundred feet of depth. The difficulties of sinking shafts and placing boilers at depths of, say, twelve thousand feet, corresponding to an increase in temperature of about 120° C., are not insuperable, and we could certainly avail ourselves in this way of the internal heat of the globe. In fact, it would not be necessary to go to any depth at all in order to derive energy from the stored terrestrial heat. The superficial layers of the earth and the air strata close to the same are at a temperature sufficiently high to evaporate some extremely volatile substances, which we might use in our boilers instead of water. There is no doubt that a vessel might be propelled on the ocean by an engine driven by such a volatile fluid, no other energy being used but the heat abstracted from the water. But the amount of power which could be obtained in this manner would be, without further provision, very small.

Electricity produced by natural causes is another source of energy which might be rendered available. Lightning discharges involve great amounts of electrical energy, which we could utilize by transforming and storing it. Some years ago I made known a method of electrical transformation which renders the first part of this task easy, but the storing of the energy of lightning discharges will be difficult to accomplish. It is well known, furthermore, that electric currents circulate constantly through the earth, and that there exists between the earth and any air stratum a difference of electrical pressure, which varies in proportion to the height.

In recent experiments I have discovered two novel facts of importance in this connection. One of these facts is that an electric current is generated in a wire extending from the ground to a great height by the axial, and probably also by the translatory, movement of the earth. No appreciable current, however, will flow continuously in the wire unless the electricity is allowed to leak out into the air. Its escape is greatly facilitated by providing at the elevated end of the wire a conducting terminal of great surface, with many sharp edges or points. We are thus enabled to get a continuous supply of electrical energy by merely supporting a wire at a height, but, unfortunately, the amount of electricity which can be so obtained is small.

The second fact which I have ascertained is that the upper air strata are permanently charged with electricity opposite to that of the earth. So, at least, I have interpreted my observations, from which it appears that the earth, with its adjacent insulating and outer conducting envelope, constitutes a highly charged electrical condenser containing, in all probability, a great amount of electrical energy which might be turned to the uses of man, if it were possible to reach with a wire to great altitudes.

It is possible, and even probable, that there will be, in time, other resources of energy opened up, of which we have no knowledge now. We may even find ways of applying forces such as magnetism or gravity for driving machinery without using any other means. Such realizations, though highly improbable, are not impossible. An example will best convey an idea of what we can hope to attain and what we can never attain. Imagine a disk of some homogeneous material turned perfectly true and arranged to turn in frictionless bearings on a horizontal shaft above the ground. This disk, being under the above conditions perfectly balanced, would rest in any position. Now, it is possible that we may learn how to make such a disk rotate continuously and perform work by the force of gravity without any further effort on our part; but it is perfectly impossible for the disk to turn and to do work without any force from the outside. If it could do so, it would be what is designated scientifically as a "perpetuum mobile," a machine creating its own motive power. To make the disk rotate by the force of gravity we have only to invent a screen against this force. By such a screen we could prevent this force from acting on one half of the disk, and the rotation of the latter would follow. At least, we cannot deny such a possibility until we know exactly the nature of the force of gravity. Suppose that this force were due to a movement comparable to that of a stream of air passing from above toward the center of the earth. The effect of such a stream upon both halves of the disk would be equal, and the latter would not rotate ordinarily; but if one half should be guarded by a plate arresting the movement, then it would turn.

A DEPARTURE FROM KNOWN METHODS — POSSIBILITY OF A "SELF-ACTING" ENGINE OR MACHINE, INANIMATE, YET CAPABLE, LIKE A LIVING BEING, OF DERIVING ENERGY FROM THE MEDIUM — THE IDEAL WAY OF OBTAINING MOTIVE POWER.

When I began the investigation of the subject under consideration, and when the preceding or similar ideas presented themselves to me for the first time, though I was then unacquainted with a number of the facts mentioned, a survey of the various ways of utilizing the energy of the medium convinced me, nevertheless, that to arrive at a thoroughly satisfactory practical solution a radical departure from the methods then known had to be made. The windmill, the solar engine, the engine driven by terrestrial heat, had their limitations in the amount of power obtainable. Some new way had to be discovered which would enable us to get more energy. There was enough heat-energy in the medium, but only a small part of it was available for the operation of an engine in the ways then known. Besides, the energy was obtainable only at a very slow rate. Clearly, then, the problem was to discover some new method which would make it possible both to utilize more of the heat-energy of the medium and also to draw it away from the same at a more rapid rate.

I was vainly endeavoring to form an idea of how this might be accomplished, when I read some statements from Carnot and Lord Kelvin (then Sir William Thomson) which meant virtually that it is impossible for an inanimate mechanism or self-acting machine to cool a portion of the medium below the temperature of the surrounding, and operate by the

195

heat abstracted. These statements interested me intensely. Evidently a living being could do this very thing, and since the experiences of my early life which I have related had convinced me that a living being is only an automaton, or, otherwise stated, a "self-acting-engine," I came to the conclusion that it was possible to construct a machine which would do the same. As the first step toward this realization I conceived the following mechanism. Imagine a thermopile consisting of a number of bars of metal extending from the earth to the outer space beyond the atmosphere. The heat from below, conducted upward along these metal bars, would cool the earth or the sea or the air, according to the location of the lower parts of the bars, and the result, as is well known, would be an electric current circulating in these bars. The two terminals of the thermopile could now be joined through an electric motor, and, theoretically, this motor would run on and on, until the media below would be cooled down to the temperature of the outer space. This would be an inanimate engine which, to all evidence, would be cooling a portion of the medium below the temperature of the surrounding, and operating by the heat abstracted.

Diagram b. — Obtaining Energy from The Ambient Medium.

Note to Dia. b. — *A*, medium with little energy; *B, B,* ambient medium with much energy; *O*, path of the energy.

But was it not possible to realize a similar condition without necessarily going to a height? Conceive, for the sake of illustration, an inclosure *T*, as illustrated in diagram b, such that energy could not be transferred across it except through a channel or path *O*, and that, by some means or other, in this inclosure a medium were maintained which would have little energy, and that on the outer side of the same there would be the ordinary ambient medium with much energy. Under these assumptions the energy would flow through the path *O*, as indicated by the arrow, and might then be converted on its passage into some other form of energy. The question was, Could such a condition be attained? Could we produce artificially such a "sink" for the energy of the ambient medium to flow in? Suppose that an extremely low temperature could be maintained by some process in a given space; the surrounding medium would then be compelled to give off heat, which could be converted into mechanical or other form of energy, and utilized. By realizing such a plan, we should be enabled to get at any point of the globe a continuous supply of energy, day and night. More than this, reasoning in the abstract, it would seem possible to cause a quick circulation of the medium, and thus draw the energy at a very rapid rate.

Here, then, was an idea which, if realizable, afforded a happy solution of the problem of getting energy from the medium. But was it realizable? I convinced myself that it was so in a number of ways, of which one is the following. As regards heat, we are at a high level, which may be represented by the surface of a mountain lake considerably above the sea, the level of which may mark the absolute zero of temperature existing in the interstellar space. Heat, like water, flows from high to low level, and, consequently, just as we can let the water of the lake run down to the sea, so we are able to let heat from the earth's surface travel up into the cold region above. Heat, like water, can perform work in flowing down, and if we had any doubt as to whether we could derive energy from the medium by means of a thermopile, as before described, it would be dispelled by this analogue. But can we produce cold in a given portion of the space and cause the heat to flow in continually? To create such a "sink," or "cold hole," as we might say, in the medium, would be equivalent to producing in the lake a space either empty or filled with something much lighter than water. This we could do by placing in the lake a tank, and pumping all the water out of the latter. We know, then, that the water, if allowed to flow back into the tank, would, theoretically, be able to perform exactly the same amount of work which was used in pumping it out, but not a bit more. Consequently nothing could be gained in this double operation of first raising the water and then letting it fall down. This would mean that it is impossible to create such a sink in the medium. But let us reflect a moment. Heat, though following certain general laws of mechanics, like a fluid, is not such; it is energy which may be converted into other forms of energy as it passes from a high to a low level. To make our mechanical analogy complete and true, we must, therefore, assume that the water, in its passage into the tank, is converted into something else, which may be taken out of it without using any, or by using very little, power. For example, if heat be represented in this analogue by the water of the lake, the oxygen and hydrogen composing the water may illustrate other forms of energy into which the heat is transformed in passing from hot to cold. If the process of heat transformation were absolutely perfect, no heat at all would arrive at the low level, since all of it would be converted into other forms of energy.

Corresponding to this ideal case, all the water flowing into the tank would be decomposed into oxygen and hydrogen before reaching the bottom, and the result would be that water would continually flow in, and yet the tank would remain entirely empty, the gases formed escaping. We would thus produce, by expending initially a certain amount of work to create a sink for the heat or, respectively, the water to flow in, a condition enabling us to get any amount of energy without further effort. This would be an ideal way of obtaining motive power. We do not know of any such absolutely perfect process of heat-conversion, and consequently some heat will generally reach the low level, which means to say, in our mechanical analogue, that some water will arrive at the bottom of the tank, and a gradual and slow filling of the latter will take place, necessitating continuous pumping out. But evidently there will be less to pump out than flows in, or, in other words, less energy will be needed to maintain the initial condition than is developed by the fall, and this is to say that some energy will be gained from the medium. What is not converted in flowing down can just be raised up with its own energy, and what is converted is clear gain. Thus the virtue of the principle I have discovered resides wholly in the conversion of the energy on the downward flow.

FIRST EFFORTS TO PRODUCE THE SELF-ACTING ENGINE — THE MECHANICAL OSCILLATOR — WORK OF DEWAR AND LINDE — LIQUID AIR.

Having recognized this truth, I began to devise means for carrying out my idea, and, after long thought, I finally conceived a combination of apparatus which should make possible the obtaining of power from the medium by a process of continuous cooling of atmospheric air. This apparatus, by continually transforming heat into mechanical work, tended to become colder and colder, and if it only were practicable to reach a very low temperature in this manner, then a sink for the heat could be produced, and energy could be derived from the medium. This seemed to be contrary to the statements of Carnot and Lord Kelvin before referred to, but I concluded from the theory of the process that such a result could be attained. This conclusion I reached, I think, in the latter part of 1883, when I was in Paris, and it was at a time when my mind was being more and more dominated by an invention which I had evolved during the preceding year, and which has since become known under the name of the "rotating magnetic field." During the few years which followed I elaborated further the plan I had imagined, and studied the working conditions, but made little headway. The commercial introduction in this country of the invention before referred to required most of my energies until 1889, when I again took up the idea of the self-acting machine. A closer investigation of the principles involved, and calculation, now showed that the result I aimed at could not be reached in a practical manner by ordinary machinery, as I had in the beginning expected. This led me, as a next step, to the study of a type of engine generally designated as "turbine," which at first seemed to offer better chances for a realization of the idea. Soon I found, however, that the turbine, too, was unsuitable. But my conclusions showed that if an engine of a peculiar kind could be brought to a high degree of perfection, the plan I had conceived was realizable, and I resolved to proceed with the development of such an engine, the primary object of which was to secure the greatest economy of transformation of heat into mechanical energy. A characteristic feature of the engine was that the work-performing piston was not connected with anything else, but was perfectly free to vibrate at an enormous rate. The mechanical difficulties encountered in the construction of this engine were greater than I had anticipated, and I made slow progress. This work was continued until early in 1892, when I went to London, where I saw Professor Dewar's admirable experiments with liquefied gases. Others had liquefied gases before, and notably Ozlewski and Pictet had performed creditable early experiments in this line, but there was such a vigor about the work of Dewar that even the old appeared new. His experiments showed, though in a way different from that I had imagined, that it was possible to reach a very low temperature by transforming heat into mechanical work, and I returned, deeply impressed with what I had seen, and more than ever convinced that my plan was practicable. The work temporarily interrupted was taken up anew, and soon I had in a fair state of perfection the engine which I have named "the mechanical oscillator." In this machine I succeeded in doing away with all packings, valves, and lubrication, and in producing so rapid a vibration of the piston that shafts of tough steel, fastened to the same and vibrated longitudinally, were torn asunder. By combining this engine with a dynamo of special design I produced a highly efficient electrical generator, invaluable in measurements and determinations of physical quantities on account of the unvarying rate of oscillation obtainable by its means. I exhibited several types of this machine, named "mechanical and electrical oscillator," before the Electrical Congress at the World's Fair in Chicago during the summer of 1893, in a lecture which, on account of other pressing work, I was unable to prepare for publication. On that occasion I exposed the principles of the mechanical oscillator, but the original purpose of this machine is explained here for the first time.

In the process, as I had primarily conceived it, for the utilization of the energy of the ambient medium, there were five essential elements in combination, and each of these had to be newly designed and perfected, as no such machines existed. The mechanical oscillator was the first element of this combination, and having perfected this, I turned to the next, which was an air-compressor of a design in certain respects resembling that of the mechanical oscillator. Similar difficulties in the construction were again encountered, but the work was pushed vigorously, and at the close of 1894 I had completed these two elements of the combination, and thus produced an apparatus for compressing air, virtually to any desired pressure, incomparably simpler, smaller, and more efficient than the ordinary. I was just beginning work on the third element, which together with the first two would give a refrigerating machine of exceptional efficiency and simplicity, when a misfortune befell me in the burning of my laboratory, which crippled my labors and delayed me. Shortly afterward Dr. Carl Linde announced the liquefaction of air by a self-cooling process, demonstrating that it was practicable to proceed with the cooling until liquefaction of the air took place. This was the only experimental proof which I was still wanting that energy was obtainable from the medium in the manner contemplated by me.

The liquefaction of air by a self-cooling process was not, as popularly believed, an accidental discovery, but a scientific result which could not have been delayed much longer, and which, in all probability, could not have escaped Dewar. This fascinating advance, I believe, is largely due to the powerful work of this great Scotchman. Nevertheless, Linde's is an immortal achievement. The manufacture of liquid air has been carried on for four years in Germany, on a scale much larger than in any other country, and this strange product has been applied for a variety of purposes. Much was expected of it in the beginning, but so far it has been an industrial ignis fatuus. By the use of such machinery as I am perfecting, its cost will probably be greatly lessened, but even then its commercial success will be questionable. When used as a refrigerant it is uneconomical, as its temperature is unnecessarily low. It is as expensive to maintain a body at a very low temperature as it is to keep it very hot; it takes coal to keep air cold. In oxygen manufacture it cannot yet compete with the electrolytic method. For use as an explosive it is unsuitable, because its low temperature again condemns it to a small efficiency, and for motive-power purposes its cost is still by far too high. It is of interest to note, however, that in driving an engine by liquid air a certain amount of energy may be gained from the engine, or, stated otherwise, from the ambient medium which keeps the engine warm, each two hundred pounds of iron-casting of the latter contributing energy at the rate of about one effective horse-power during one hour. But this gain of the consumer is offset by an equal loss of the producer.

Much of this task on which I have labored so long remains to be done. A number of mechanical details are still to be perfected and some difficulties of a different nature to be mastered, and I cannot hope to produce a self-acting machine deriving energy from the ambient medium for a long time yet, even if all my expectations should materialize. Many circumstances have occurred which have retarded my work of late, but for several reasons the delay was beneficial.

One of these reasons was that I had ample time to consider what the ultimate possibilities of this development might be. I worked for a long time fully convinced that the practical realization of this method of obtaining energy from the sun would be of incalculable industrial value, but the continued study of the subject revealed the fact that while it will be commercially profitable if my expectations are well founded, it will not be so to an extraordinary degree.

DISCOVERY OF UNEXPECTED PROPERTIES OF THE ATMOSPHERE — STRANGE EXPERIMENTS — TRANSMISSION OF ELECTRICAL ENERGY THROUGH ONE WIRE WITHOUT RETURN — TRANSMISSION THROUGH THE EARTH WITHOUT ANY WIRE.

Another of these reasons was that I was led to recognize the transmission of electrical energy to any distance through the media as by far the best solution of the great problem of harnessing the sun's energy for the uses of man. For a long time I was convinced that such a transmission on an industrial scale could never be realized, but a discovery which I made changed my view. I observed that under certain conditions the atmosphere, which is normally a high insulator, assumes conducting properties, and so becomes capable of conveying any amount of electrical energy. But the difficulties in the way of a practical utilization of this discovery for the purpose of transmitting electrical energy without wires were seemingly insuperable. Electrical pressures of many millions of volts had to be produced and handled; generating apparatus of a novel kind, capable of withstanding the immense electrical stresses, had to be invented and perfected, and a complete safety against the dangers of the high-tension currents had to be attained in the system before its practical introduction could be even thought of. All this could not be done in a few weeks or months, or even years. The work required patience and constant application, but the improvements came, though slowly. Other valuable results were, however, arrived at in the course of this long-continued work, of which I shall endeavor to give a brief account, enumerating the chief advances as they were successively effected.

The discovery of the conducting properties of the air, though unexpected, was only a natural result of experiments in a special field which I had carried on for some years before. It was, I believe, during 1889 that certain possibilities offered by extremely rapid electrical oscillations determined me to design a number of special machines adapted for their investigation. Owing to the peculiar requirements, the construction of these machines was very difficult, and consumed much time and effort; but my work on them was generously rewarded, for I reached by their means several novel and important results. One of the earliest observations I made with these new machines was that electrical oscillations of an extremely high rate act in an extraordinary manner upon the human organism. Thus, for instance, I demonstrated that powerful electrical discharges of several hundred thousand volts, which at that time were considered absolutely deadly, could be passed through the body without inconvenience or hurtful consequences. These oscillations produced other specific physiological effects, which, upon my announcement, were eagerly taken up by skilled physicians and further investigated. This new field has proved itself fruitful beyond expectation, and in the few years which have passed since, it has been developed to such an extent that it now forms a legitimate and important department of medical science. Many results, thought impossible at that time, are now readily obtainable with these oscillations, and many experiments undreamed of then can now be readily performed by their means. I still remember with pleasure how, nine years ago, I passed the discharge of a powerful induction-coil through my body to demonstrate before a scientific society the comparative harmlessness of very rapidly vibrating electric currents, and I can still recall the astonishment of my audience. I would now undertake, with much less apprehension than I had in that experiment, to transmit through my body with such currents the entire electrical energy of the dynamos now working at Niagara — forty or fifty thousand horse-power. I have produced electrical oscillations which were of such intensity that when circulating through my arms and chest they have melted wires which joined my hands, and still I felt no inconvenience. I have energized with such oscillations a loop of heavy copper wire so powerfully that masses of metal, and even objects of an electrical resistance specifically greater than that of human tissue, brought close to or placed within the loop, were heated to a high temperature and melted, often with the violence of an explosion, and yet into this very space in which this terribly destructive turmoil was going on I have repeatedly thrust my head without feeling anything or experiencing injurious after-effects.

Another observation was that by means of such oscillations light could be produced in a novel and more economical manner, which promised to lead to an ideal system of electric illumination by vacuum-tubes, dispensing with the necessity of renewal of lamps or incandescent filaments, and possibly also with the use of wires in the interior of buildings. The efficiency of this light increases in proportion to the rate of the oscillations, and its commercial success is, therefore, dependent on the economical production of electrical vibrations of transcending rates. In this direction I have met with gratifying success of late, and the practical introduction of this new system of illumination is not far off.

The investigations led to many other valuable observations and results, one of the more important of which was the demonstration of the practicability of supplying electrical energy through one wire without return. At first I was able to transmit in this novel manner only very small amounts of electrical energy, but in this line also my efforts have been rewarded with similar success.

The photograph shown in Fig. 3 (see p. 186) illustrates, as its title explains, an actual transmission of this kind effected with apparatus used in other experiments here described. To what a degree the appliances have been perfected since my first demonstrations early in 1891 before a scientific society, when my apparatus was barely capable of lighting one lamp (which result was considered wonderful), will appear when I state that I have now no difficulty in lighting in this manner four or five hundred lamps, and could light many more. In fact, there is no limit to the amount of energy which may in this way be supplied to operate any kind of electrical device.

After demonstrating the practicability of this method of transmission, the thought naturally occurred to me to use the earth as a conductor, thus dispensing with all wires. Whatever electricity may be, it is a fact that it behaves like an incompressible fluid, and the earth may be looked upon as an immense reservoir of electricity, which, I thought, could be disturbed effectively by a properly designed electrical machine. Accordingly, my next efforts were directed toward perfecting a special apparatus which would be highly effective in creating a disturbance of electricity in the earth. The progress in this new direction was necessarily very slow and the work discouraging, until I finally succeeded in perfecting a novel kind of transformer or induction-coil, particularly suited for this special purpose. That it is practicable, in this manner, not only to transmit minute amounts of electrical energy for operating delicate electrical devices, as I contemplated at first, but also electrical energy in appreciable quantities, will appear from an inspection of Fig. 4 (see p. 186), which illustrates an actual experiment of this kind performed with the same apparatus. The result obtained was all the more remarkable as the top end of the coil was not connected to a wire or plate for magnifying the effect.

"WIRELESS" TELEGRAPHY — THE SECRET OF TUNING — ERRORS IN THE HERTZIAN INVESTIGATIONS — A RECEIVER OF WONDERFUL SENSITIVENESS.

Diagram c. — "Wireless" Telegraphy Mechanically Illustrated.

As the first valuable result of my experiments in this latter line a system of telegraphy without wires resulted, which I described in two scientific lectures in February and March, 1893. It is mechanically illustrated in diagram c, the upper part of which shows the electrical arrangement as I described it then, while the lower part illustrates its mechanical analogue. The system is extremely simple in principle. Imagine two tuning-forks F, F_1, one at the sending- and the other at the receiving-station respectively, each having attached to its lower prong a minute piston p, fitting in a cylinder. Both the cylinders communicate with a large reservoir R, with elastic walls, which is supposed to be closed and filled with a light and incompressible fluid. By striking repeatedly one of the prongs of the tuning-fork F, the small piston p below would be vibrated, and its vibrations, transmitted through the fluid, would reach the distant fork F_1, which is "tuned" to the fork F, or, stated otherwise, of exactly the same note as the latter. The fork F_1 would now be set vibrating, and its vibration would be intensified by the continued action of the distant fork F until its upper prong, swinging far out, would make an electrical connection with a stationary contact c'', starting in this manner some electrical or other appliances which may be used for recording the signals. In this simple way messages could be exchanged between the two stations, a similar contact c' being provided for this purpose, close to the upper prong of the fork F, so that the apparatus at each station could be employed in turn as receiver and transmitter.

The electrical system illustrated in the upper figure of diagram c is exactly the same in principle, the two wires or circuits ESP and $E_1S_1P_1$, which extend vertically to a height, representing the two tuning-forks with the pistons attached to them. These circuits are connected with the ground by plates E, E_1, and to two elevated metal sheets P, P_1, which store electricity and thus magnify considerably the effect. The closed reservoir R, with elastic walls, is in this case replaced by the earth, and the fluid by electricity. Both of these circuits are "tuned" and operate just like the two tuning-forks. Instead of striking the fork F at the sending-station, electrical oscillations are produced in the vertical sending- or transmitting-wire ESP, as by the action of a source S, included in this wire, which spread through the ground and reach the distant vertical receiving-wire $E_1S_1P_1$, exciting corresponding electrical oscillations in the same. In the latter wire or circuit is included a sensitive device or receiver S_1, which is thus set in action and made to operate a relay or other appliance. Each station is, of course, provided both with a source of electrical oscillations S and a sensitive receiver S_1, and a simple provision is made for using each of the two wires alternately to send and to receive the messages.

The exact attunement of the two circuits secures great advantages, and, in fact, it is essential in the practical use of the system. In this respect many popular errors exist, and, as a rule, in the technical reports on this subject circuits and appliances are described as affording these advantages when from their very nature it is evident that this is impossible. In order to attain the best results it is essential that the length of each wire or circuit, from the ground connection to the top, should be equal to one quarter of the wave-length of the electrical vibration in the wire, or else equal to that length multiplied by an odd number. Without the observation of this rule it is virtually impossible to prevent the interference and insure the privacy of messages. Therein lies the secret of tuning. To obtain the most satisfactory results it is, however, necessary to resort to electrical vibrations of low pitch. The Hertzian spark apparatus, used generally by experimenters, which produces oscillations of a very high rate, permits no effective tuning, and slight disturbances are sufficient to render an exchange of messages impracticable. But scientifically designed, efficient appliances allow nearly perfect adjustment. An experiment performed with the improved apparatus repeatedly referred to, and intended to convey an idea of this feature, is illustrated in Fig. 5 (p. 187), which is sufficiently explained by its note.

Since I described these simple principles of telegraphy without wires I have had frequent occasion to note that the identical features and elements have been used, in the evident belief that the signals are being transmitted to considerable distance by "Hertzian" radiations. This is only one of many misapprehensions to which the investigations of the lamented physicist have given rise. About thirty-three years ago Maxwell, following up a suggestive experiment made by Faraday in 1845, evolved an ideally simple theory which intimately connected light, radiant heat, and electrical phenomena, interpreting them as being all due to vibrations of a hypothetical fluid of inconceivable tenuity, called the ether. No experimental verification was arrived at until Hertz, at the suggestion of Helmholtz, undertook a series of experiments to this effect. Hertz proceeded with extraordinary ingenuity and insight, but devoted little energy to the perfection of his old-fashioned apparatus. The consequence was that he failed to observe the important function which the air played in his experiments, and which I subsequently discovered. Repeating his experiments and reaching different results, I ventured to point out this oversight. The strength of the proofs brought forward by Hertz in support of Maxwell's theory resided in the correct estimate of the rates of vibration of the circuits he used. But I ascertained that he could not have obtained the rates he thought he was getting. The vibrations with identical apparatus he employed are, as a rule, much slower, this being due to the presence of air, which produces a dampening effect upon a rapidly vibrating electric circuit of high pressure, as a fluid does upon a vibrating tuning-fork. I have, however, discovered since that time other causes of error, and I have long ago ceased to look upon his results as being an experimental verification of the poetical conceptions of Maxwell. The work of the great German physicist has acted as an immense stimulus to contemporary electrical research, but it has likewise, in a measure, by its fascination, paralyzed the scientific mind, and thus hampered independent inquiry. Every new phenomenon which was discovered was made to fit the theory, and so very often the truth has been unconsciously distorted.

When I advanced this system of telegraphy, my mind was dominated by the idea of effecting communication to any distance through the earth or environing medium, the practical consummation of which I considered of transcendent importance, chiefly on account of the moral effect which it could not fail to produce universally. As the first effort to this end I proposed at that time, to employ relay-stations with tuned circuits, in the hope of making thus practicable signaling over vast distances, even with apparatus of very moderate power then at my command. I was confident, however, that with properly designed machinery signals could be transmitted to any point of the globe, no matter what the distance, without the necessity of using such intermediate stations. I gained this conviction through the discovery of a singular electrical phenomenon, which I described early in 1892, in lectures I delivered before some scientific societies abroad, and which I have called a "rotating brush." This is a bundle of light which is formed, under certain conditions, in a vacuum-bulb, and which is of a sensitiveness to magnetic and electric influences bordering, so to speak, on the supernatural. This light-bundle is rapidly rotated by the earth's magnetism as many as twenty thousand times per second, the rotation in these parts being opposite to what it would be in the southern hemisphere, while in the region of the magnetic equator it should not rotate at all. In its most sensitive state, which is difficult to attain, it is responsive to electric or magnetic influences to an incredible degree. The mere stiffening of the muscles of the arm and consequent slight electrical change in the body of an observer standing at some distance from it, will perceptibly affect it. When in this highly sensitive state it is capable of indicating the slightest magnetic and electric changes taking place in the earth. The observation of this wonderful phenomenon impressed me strongly that communication at any distance could be easily effected by its means, provided that apparatus could be perfected capable of producing an electric or magnetic change of state, however small, in the terrestrial globe or environing medium.

DEVELOPMENT OF A NEW PRINCIPLE — THE ELECTRICAL OSCILLATOR — PRODUCTION OF IMMENSE ELECTRICAL MOVEMENTS — THE EARTH RESPONDS TO MAN — INTERPLANETARY COMMUNICATION NOW PROBABLE.

I resolved to concentrate my efforts upon this venturesome task, though it involved great sacrifice, for the difficulties to be mastered were such that I could hope to consummate it only after years of labor. It meant delay of other work to which I would have preferred to devote myself, but I gained the conviction that my energies could not be more

213

usefully employed; for I recognized that an efficient apparatus for the production of powerful electrical oscillations, as was needed for that specific purpose, was the key to the solution of other most important electrical and, in fact, human problems. Not only was communication, to any distance, without wires possible by its means, but, likewise, the transmission of energy in great amounts, the burning of the atmospheric nitrogen, the production of an efficient illuminant, and many other results of inestimable scientific and industrial value. Finally, however, I had the satisfaction of accomplishing the task undertaken by the use of a new principle, the virtue of which is based on the marvelous properties of the electrical condenser. One of these is that it can discharge or explode its stored energy in an inconceivably short time. Owing to this it is unequaled in explosive violence. The explosion of dynamite is only the breath of a consumptive compared with its discharge. It is the means of producing the strongest current, the highest electrical pressure, the greatest commotion in the medium. Another of its properties, equally valuable, is that its discharge may vibrate at any rate desired up to many millions per second.

I had arrived at the limit of rates obtainable in other ways when the happy idea presented itself to me to resort to the condenser. I arranged such an instrument so as to be charged and discharged alternately in rapid succession through a coil with a few turns of stout wire, forming the primary of a transformer or induction-coil. Each time the condenser was discharged the current would quiver in the primary wire and induce corresponding oscillations in the secondary. Thus a transformer or induction-coil on new principles was evolved, which I have called "the electrical oscillator," partaking of those unique qualities which characterize the condenser, and enabling results to be attained impossible by other means. Electrical effects of any desired character and of intensities undreamed of before are now easily producible by perfected apparatus of this kind, to which frequent reference has been made, and the essential parts of which are shown in Fig. 6 (p. 188). For certain purposes a strong inductive effect is required; for others the greatest possible suddenness; for others again, an exceptionally high rate of vibration or extreme pressure; while for certain other objects immense electrical movements are necessary. The photographs in Figs. 7, 8, 9, and 10, of experiments performed with such an oscillator, may serve to illustrate some of these features and convey an idea of the magnitude of the effects actually produced. The completeness of the titles of the figures referred to makes a further description of them unnecessary.

However extraordinary the results shown may appear, they are but trifling compared with those which are attainable by apparatus designed on these same principles. I have produced electrical discharges the actual path of which, from end to end, was probably more than one hundred feet long; but it would not be difficult to reach lengths one hundred times as great. I have produced electrical movements occurring at the rate of approximately one hundred thousand horse-power, but rates of one, five, or ten million horse-power are easily practicable. In these experiments effects were developed incomparably greater than any ever produced by human agencies, and yet these results are but an embryo of what is to be.

That communication without wires to any point of the globe is practicable with such apparatus would need no demonstration, but through a discovery which I made I obtained absolute certitude. Popularly explained, it is exactly this: When we raise the voice and hear an echo in reply, we know that the sound of the voice must have reached a distant wall, or boundary, and must have been reflected from the same. Exactly as the sound, so an electrical wave is reflected, and the same evidence which is afforded by an echo is offered by an electrical phenomenon known as a "stationary" wave — that is, a wave with fixed nodal and ventral regions. Instead of sending sound-vibrations toward a distant wall, I have sent electrical vibrations toward the remote boundaries of the earth, and instead of the wall the earth has replied. In place of an echo I have obtained a stationary electrical wave, a wave reflected from afar.

Stationary waves in the earth mean something more than mere telegraphy without wires to any distance. They will enable us to attain many important specific results impossible otherwise. For instance, by their use we may produce at will, from a sending-station, an electrical effect in any particular region of the globe; we may determine the relative position or course of a moving object, such as a vessel at sea, the distance traversed by the same, or its speed; or we may send over the earth a wave of electricity traveling at any rate we desire, from the pace of a turtle up to lightning speed.

With these developments we have every reason to anticipate that in a time not very distant most telegraphic messages across the oceans will be transmitted without cables. For short distances we need a "wireless" telephone, which requires no expert operators. The greater the spaces to be bridged, the more rational becomes communication without wires. The cable is not only an easily damaged and costly instrument, but it limits us in the speed of transmission by reason of a certain electrical property inseparable from its construction. A properly designed plant for effecting communication without wires ought to have many times the working capacity of a cable, while it will involve incomparably less expense. Not a long time will pass, I believe, before communication by cable will become obsolete, for not only will signaling by this new method be quicker and cheaper, but also much safer. By using some new means for isolating the messages which I have contrived, an almost perfect privacy can be secured.

I have observed the above effects so far only up to a limited distance of about six hundred miles, but inasmuch as there is virtually no limit to the power of the vibrations producible with such an oscillator, I feel quite confident of the success of such a plant for effecting transoceanic communication. Nor is this all. My measurements and calculations have shown that it is perfectly practicable to produce on our globe, by the use of these principles, an electrical movement of such magnitude that, without the slightest doubt, its effect will be perceptible on some of our nearer planets, as Venus and Mars. Thus from mere possibility interplanetary communication has entered the stage of probability. In fact, that we can produce a distinct effect on one of these planets in this novel manner, namely, by disturbing the electrical condition of the earth, is beyond any doubt. This way of effecting such communication is, however, essentially different from all others which have so far been proposed by scientific men. In all the previous instances only a minute fraction of the total energy reaching the planet — as much as it would be possible to concentrate in a reflector — could be utilized by the supposed observer in his instrument. But by the means I have developed he would be enabled to concentrate the larger portion of the entire energy transmitted to the planet in his instrument, and the chances of affecting the latter are thereby increased many millionfold.

Besides machinery for producing vibrations of the required power, we must have delicate means capable of revealing the effects of feeble influences exerted upon the earth. For such purposes, too, I have perfected new methods. By their use we shall likewise be able, among other things, to detect at considerable distance the presence of an iceberg or other object at sea. By their use, also, I have discovered some terrestrial phenomena still unexplained. That we can send a message to a planet is certain, that we can get an answer is probable: man is not the only being in the Infinite gifted with a mind.

TRANSMISSION OF ELECTRICAL ENERGY TO ANY DISTANCE WITHOUT WIRES — NOW PRACTICABLE — THE BEST MEANS OF INCREASING THE FORCE ACCELERATING THE HUMAN MASS.

The most valuable observation made in the course of these investigations was the extraordinary behavior of the atmosphere toward electric impulses of excessive electromotive force. The experiments showed that the air at the ordinary pressure became distinctly conducting, and this opened up the wonderful prospect of transmitting large amounts of electrical energy for industrial purposes to great distances without wires, a possibility which, up to that time, was thought of only as a scientific dream. Further investigation revealed the important fact that the conductivity imparted to the air by these electrical impulses of many millions of volts increased very rapidly with the degree of rarefaction, so that air strata at very moderate altitudes, which are easily accessible, offer, to all experimental evidence, a perfect conducting path, better than a copper wire, for currents of this character.

Thus the discovery of these new properties of the atmosphere not only opened up the possibility of transmitting, without wires, energy in large amounts, but, what was still more significant, it afforded the certitude that energy could be transmitted in this manner economically. In this new system it matters little — in fact, almost nothing — whether the transmission is effected at a distance of a few miles or of a few thousand miles.

While I have not, as yet, actually effected a transmission of a considerable amount of energy, such as would be of industrial importance, to a great distance by this new method, I have operated several model plants under exactly the same conditions which will exist in a large plant of this kind, and the practicability of the system is thoroughly demonstrated. The experiments have shown conclusively that, with two terminals maintained at an elevation of not more than thirty thousand to thirty-five thousand feet above sea-level, and with an electrical pressure of fifteen to twenty million volts, the energy of thousands of horse-power can be transmitted over distances which may be hundreds and, if necessary, thousands of miles. I am hopeful, however, that I may be able to reduce very considerably the elevation of the terminals now required, and with this object I am following up an idea which promises such a realization. There is, of course, a popular prejudice against using an electrical pressure of millions of volts, which may cause sparks to fly at distances of hundreds of feet, but, paradoxical as it may seem, the system, as I have described it in a technical publication, offers greater personal safety than most of the ordinary distribution circuits now used in the cities. This is, in a measure, borne out by the fact that, although I have carried on such experiments for a number of years, no injury has been sustained either by me or any of my assistants.

But to enable a practical introduction of the system, a number of essential requirements are still to be fulfilled. It is not enough to develop appliances by means of which such a transmission can be effected. The machinery must be such as to allow the transformation and transmission of electrical energy under highly economical and practical conditions. Furthermore, an inducement must be offered to those who are engaged in the industrial exploitation of natural sources of power, as waterfalls, by guaranteeing greater returns on the capital invested than they can secure by local development of the property.

From that moment when it was observed that, contrary to the established opinion, low and easily accessible strata of the atmosphere are capable of conducting electricity, the transmission of electrical energy without wires has become a rational task of the engineer, and one surpassing all others in importance. Its practical consummation would mean that energy would be available for the uses of man at any point of the globe, not in small amounts such as might be derived from the ambient medium by suitable machinery, but in quantities virtually unlimited, from waterfalls. Export of power would then become the chief source of income for many happily situated countries, as the United States, Canada, Central and South America, Switzerland, and Sweden. Men could settle down everywhere, fertilize and irrigate the soil with little effort, and convert barren deserts into gardens, and thus the entire globe could be transformed and made a fitter abode for mankind. It is highly probable that if there are intelligent beings on Mars they have long ago realized this very idea, which would explain the changes on its surface noted by astronomers. The atmosphere on that planet, being of considerably smaller density than that of the earth, would make the task much more easy.

It is probable that we shall soon have a self-acting heat-engine capable of deriving moderate amounts of energy from the ambient medium. There is also a possibility — though a small one — that we may obtain electrical energy direct from the sun. This might be the case if the Maxwellian theory is true, according to which electrical vibrations of all rates should emanate from the sun. I am still investigating this subject. Sir William Crookes has shown in his beautiful invention known as the "radiometer" that rays may produce by impact a mechanical effect, and this may lead to some important revelation as to the utilization of the sun's rays in novel ways. Other sources of energy may be opened up, and new methods of deriving energy from the sun discovered, but none of these or similar achievements would equal in importance the transmission of power to any distance through the medium. I can conceive of no technical advance which would tend to unite the various elements of humanity more effectively than this one, or of one which would more add to and more economize human energy. It would be the best means of increasing the force accelerating the human mass. The mere moral influence of such a radical departure would be incalculable. On the other hand if at any point of the globe energy can be obtained in limited quantities from the ambient medium by means of a self-acting heat-engine or otherwise, the conditions will remain the same as before. Human performance will be increased, but men will remain strangers as they were.

I anticipate that many, unprepared for these results, which, through long familiarity, appear to me simple and obvious, will consider them still far from practical application. Such reserve, and even opposition, of some is as useful a quality and as necessary an element in human progress as the quick receptivity and enthusiasm of others. Thus, a mass which resists the force at first, once set in movement, adds to the energy. The scientific man does not aim at an immediate result. He does not expect that his advanced ideas will be readily taken up. His work is like that of the planter — for the future. His duty is to lay the foundation for those who are to come, and point the way. He lives and labors and hopes with the poet who says:

Schaff', das Tagwerk meiner Hände,
Hohes Glück, dass ich's vollende!
Lass, o lass mich nicht ermatten!
Nein, es sind nicht leere Träume:
Jetzt nur Stangen, diese Bäume
Geben einst noch Frucht und Schatten.[1]

[1] Daily work — my hands' employment,
To complete is pure enjoyment!
Let, oh, let me never falter!
No! there is no empty dreaming:
Lo! these trees, but bare poles seeming,
Yet will yield both food and shelter!

Goethe's "Hope"
Translated by William Gibson, Com. U. S. N.

ON LIGHT AND OTHER HIGH FREQUENCY PHENOMENA

Presentations

Franklin Institute, Philadelphia,
February 24th, 1893

National Electric Light Association,
St. Louis, March 1st, 1893

INTRODUCTORY — SOME THOUGHTS ON THE EYE

When we look at the world around us, on Nature, we are impressed with its beauty and grandeur. Each thing we perceive, though it may be vanishingly small, is in itself a world, that is, like the whole of the universe, matter and force governed by law, — a world, the contemplation of which fills us with feelings of wonder and irresistibly urges us to ceaseless thought and inquiry. But in all this vast world, of all objects our senses reveal to us, the most marvelous, the most appealing to our imagination, appears no doubt a highly developed organism, a thinking being. If there is anything fitted to make us admire Nature's handiwork, it is certainly this inconceivable structure, which performs its innumerable motions of obedience to external influence. To understand its workings, to get a deeper insight into this Nature's masterpiece, has ever been for thinkers a fascinating aim, and after many centuries of arduous research men have arrived at a fair understanding of the functions of its organs and senses. Again, in all the perfect harmony of its parts, of the parts which constitute the material or tangible of our being, of all its organs and senses, the eye is the most wonderful. It is the most precious, the most indispensable of our perceptive or directive organs, it is the great gateway through which all knowledge enters the mind. Of all our organs, it is the one, which is in the most intimate relation with that which we call intellect. So intimate is this relation, that it is often said, the very soul shows itself in the eye.

It can be taken as a fact, which the theory of the action of the eye implies, that for each external impression, that is, for each image produced upon the retina, the ends of the visual nerves, concerned in the conveyance of the impression to the mind, must be under a peculiar stress or in a vibratory state, It now does not seem improbable that, when by the power of thought an image is evoked, a distinct reflex action, no matter how weak, is exerted upon certain ends of the visual nerves, and therefore upon the retina. Will it ever be within human power to analyze the condition of the retina when disturbed by thought or reflex action, by the help of some optical or other means of such sensitiveness, that a clear idea of its state might be gained at any time? If this were possible, then the problem of reading one's thoughts with precision, like the characters of an open book, might be much easier to solve than many problems belonging to the domain of positive physical science, in the solution of which many, if not the majority, of scientific men implicitly believe. Helmholtz has shown that the fundi of the eye are themselves, luminous, and he was able to *see*, in total darkness, the movement of his arm by the light of his own eyes. This is one of the most remarkable experiments recorded in the history of science, and probably only a few men could satisfactorily repeat it, for it is very likely, that the luminosity of the eyes is associated with uncommon activity of the brain and great imaginative power. It is fluorescence of brain action, as it were.

Another fact having a bearing on this subject which has probably been noted by many, since it is stated in popular expressions, but which I cannot recollect to have found chronicled as a positive result of observation is, that at times, when a sudden idea or image presents itself to the intellect, there is a distinct and sometimes painful sensation of luminosity produced in the eye, observable even in broad daylight.

The saying then, that the soul shows itself in the eye, is deeply founded, and we feel that it expresses a great truth. It has a profound meaning even for one who, like a poet or artist, only following his inborn instinct or love for Nature, finds delight in aimless thoughts and in the mere contemplation of natural phenomena, but a still more profound meaning for one who, in the spirit of positive scientific investigation, seeks to ascertain the causes of the effects. It is principally the natural philosopher, the physicist, for whom the eye is the subject of the most intense admiration.

Two facts about the eye must forcibly impress the mind of the physicist, notwithstanding he may think or say that it is an imperfect optical instrument, forgetting, that the very conception of that which is perfect or seems so to him, has been gained through this same instrument. First, the eye is, as far as our positive knowledge goes, the only organ which is *directly* affected by that subtile medium, which as science teaches us, must fill all space; secondly, it is the most sensitive of our organs, incomparably more sensitive to external impressions than any other.

The organ of hearing implies the impact of ponderable bodies, the organ of smell the transference of detached material particles, and the organs of taste, and of touch or force, the direct contact, or at least some interference of ponderable matter, and this is true even in those instances of animal organisms, in which some of these organs are developed to a degree of truly marvelous perfection. This being so, it seems wonderful that the organ of sight solely should be capable of being stirred by that, which all our other organs are powerless to detect, yet which plays an essential part in all natural phenomena, which transmits all energy and sustains all motion and, that most intricate of all, life, but which has properties such that even a scientifically trained mind cannot help drawing a distinction between it and all that is called matter. Considering merely this, and the fact that the eye, by its marvelous power, widens our otherwise very narrow range of perception far beyond the limits of the small world which is our own, to embrace myriads of other worlds, suns and stars in the infinite depths of the universe, would make it justifiable to assert, that it is an organ of a higher order. Its performances are beyond comprehension. Nature as far as we know never produced anything more wonderful. We can get barely a faint idea of its prodigious power by analyzing what it does and by comparing. When ether waves impinge upon the human body, they produce the sensations of warmth or cold, pleasure or pain, or perhaps other sensations of which we are not aware, and any degree or intensity of these sensations, which degrees are infinite in number, hence an infinite number of distinct sensations. But our sense of touch, or our sense of force, cannot reveal to us these differences in degree or intensity, unless they are very great. Now we can readily conceive how an organism, such as the human, in the eternal process of evolution, or more philosophically speaking, adaptation to Nature, being constrained to the use of only the sense of touch or force, for instance, might develop this sense to such a degree of sensitiveness or perfection, that it would be capable of distinguishing the minutest differences in the temperature of a body even at some distance, to a hundredth, or thousandth, or millionth part of a degree. Yet, even this apparently impossible performance would not begin to compare with that of the eye, which is capable of distinguishing and conveying to the mind in a single instant innumerable peculiarities of the body, be it in form, or color, or other respects. This power of the eye rests upon two things, namely, the rectilinear propagation of the disturbance by which it is effected, and upon its sensitiveness. To say that the eye is sensitive is not saying anything. Compared with it, all other organs are monstrously crude. The organ of smell which guides a dog on the trail of a deer, the organ of touch or force which guides an insect in its wanderings, the organ of hearing, which is affected by the slightest disturbances of the air, are sensitive organs, to be sure, but what are they compared with the human eye! No doubt it responds to the faintest echoes or reverberations of the medium; no doubt, it brings us tidings from other worlds, infinitely remote, but in a language we cannot as yet always understand. And why not? Because we live in a medium filled with air and other gases, vapors and a dense mass of solid particles flying about. These play an important part in many phenomena; they fritter away the energy of the vibrations before they can reach the eye; they too, are the carriers of germs of destruction, they get into our lungs and other organs, clog up the channels and imperceptibly, yet inevitably, arrest the stream of life. Could we but do away with all ponderable matter in the line of sight of the telescope, it would reveal to us undreamt of marvels. Even the unaided eye, I think, would be capable of distinguishing in the pure medium, small objects at distances measured probably by hundreds or perhaps thousands of miles.

But there is something else about the eye which impresses us still more than these wonderful features which we observed, viewing it from the standpoint of a physicist, merely as an optical instrument, — something which appeals to us more than its marvelous faculty of being directly affected by the vibrations of the medium, without interference of gross matter, and more than its inconceivable sensitiveness and discerning power. It is its significance in the processes of life. No matter what one's views on nature and life may be, he must stand amazed when, for the first time in his thoughts, he realizes the importance of the eye in the physical processes and mental performances of the human organism. And how could it be otherwise, when he realizes, that the eye is the means through which the human race has acquired the entire knowledge it possesses, that it controls all our motions, more still, and our actions.

There is no way of acquiring knowledge except through the eye. What is the foundation of all philosophical systems of ancient and modern times, in fact, of all the philosophy of man? *I am I think; I think, therefore I am*. But how could I think and how would I know that I exist, if I had not the eye? For knowledge involves consciousness; consciousness involves ideas, conceptions; conceptions involve pictures or images, and images the sense of vision, and therefore the organ of sight. But how about blind men, will be asked? Yes, a blind man may depict in magnificent poems, forms and scenes from real life, from a world he physically does not see. A blind man may touch the keys of an instrument with unerring precision, may model the fastest boat, may discover and invent, calculate and construct, may do still greater wonders — but all the blind men who have done such things have descended from those who had seeing eyes. Nature may reach the same result in many ways. Like a wave in the physical world, in the infinite ocean of the medium which pervades all, so in the world of organisms, in life, an impulse started proceeds onward, at times, may be, with the speed of light, at times, again, so slowly that for ages and ages it seems to stay, passing through processes of a complexity inconceivable to men, but in all its forms, in all its stages, its energy ever and ever integrally present. A single ray of light from a distant star falling upon the eye of a tyrant in by-gone times, may have altered the course of his life, may have changed the destiny of nations, may have transformed the surface of the globe, so intricate, so inconceivably complex are the processes in Nature. In no way can we get such an overwhelming idea of the grandeur of Nature, as when we consider, that in accordance with the law of the conservation of energy, throughout the infinite, the forces are in a perfect balance, and hence the energy of a single thought may determine the motion of a Universe. It is not necessary that every individual, not even that every generation or many generations, should have the physical instrument of sight, in order to be able to form images and to think, that is, form ideas or conceptions; but sometime or other, during the process of evolution, the eye certainly must have existed, else thought, as we understand it, would be impossible; else conceptions, like spirit, intellect, mind, call it as you may, could not exist. It is conceivable, that in some other world, in some other beings, the eye is replaced by a different organ, equally or more perfect, but these beings cannot be men.

Now what prompts us all to voluntary motions and actions of any kind? Again the eye. If I am conscious of the motion, I must have an idea or conception, that is, an image, therefore the eye. If I am not precisely conscious of the motion, it is, because the images are vague or indistinct, being blurred by the superimposition of many. But when I perform the motion, does the impulse which prompts me to the action come from within or from without? The greatest physicists have not disdained to endeavor to answer this and similar questions and have at times abandoned themselves to the delights of pure and unrestrained thought. Such questions are generally considered not to belong to the realm of positive physical science, but will before long be annexed to its domain. Helmholtz has probably thought more on life than any modern scientist. Lord Kelvin expressed his belief that life's process is electrical and that there is a force inherent to the organism and determining its motions. Just as much as I am convinced of any physical truth I am convinced that the motive impulse must come from the outside. For, consider the lowest organism we know — and there are probably many lower ones — an aggregation of a few cells only. If it is capable of voluntary motion it can perform an infinite number of motions, all definite and precise. But now a mechanism consisting of a finite number of parts and few at that, cannot perform are infinite number of definite motions, hence the impulses which govern its movements must come from the environment. So, the atom, the ulterior element of the Universe's structure, is tossed about in space eternally, a play to external influences, like a boat in a troubled sea. Were it to stop its motion *it would die*. Matter at rest, if such a thing could exist, would be matter dead. Death of matter! Never has a sentence of deeper philosophical meaning been uttered. This is the way in which Prof. Dewar forcibly expresses it in the description of his admirable experiments, in which liquid oxygen is handled as one handles water, and air at ordinary pressure is made to condense and even to solidify by the intense cold. Experiments, which serve to illustrate, in his language, the last feeble manifestations of life, the last quiverings of matter about to die. But human eyes shall not witness such death. There is no death of matter, for throughout the infinite universe, all has to move, to vibrate, that is, to live.

I have made the preceding statements at the peril of treading upon metaphysical ground, in my desire to introduce the subject of this lecture in a manner not altogether uninteresting, I may hope, to an audience such as I have the honor to address. But now, then, returning to the subject, this divine organ of sight, this indispensable instrument for thought and all intellectual enjoyment, which lays open to us the marvels of this universe, through which we have acquired what knowledge we possess, and which prompts us to, and controls, all our physical and mental activity. By what is it affected? By light! What is light?

We have witnessed the great strides which have been made in all departments of science in recent years. So great have been the advances that we cannot refrain from asking ourselves, Is this all true, or is it but a dream? Centuries ago men have lived, have thought, discovered, invented, and have believed that they were soaring, while they were merely proceeding at a snail's pace. So we too may be mistaken. But taking the truth of the observed events as one of the implied facts of science, we must rejoice in the immense progress already made and still more in the anticipation of what must come, judging from the possibilities opened up by modern research. There is, however, an advance which we have been witnessing, which must be particularly gratifying to every lover of progress. It is not a discovery, or an invention, or an achievement in any particular direction. It is an advance in all directions of scientific thought and experiment. I mean the generalization of the natural forces and phenomena, the looming up of a certain broad idea on the scientific horizon. It is this idea which has, however, long ago taken possession of the most advanced minds, to which I desire to call your attention, and which I intend to illustrate in a general way, in these experiments, as the first step in answering the question "What is light?" and to realize the modern meaning of this word.

It is beyond the scope of my lecture to dwell upon the subject of light in general, my object being merely to bring presently to your notice a certain class of light effects and a number of phenomena observed in pursuing the study of these effects. But to he consistent in my remarks it is necessary to state that, according to that idea, now, accepted by the majority of scientific men as a positive result of theoretical and experimental investigation, the various forms or manifestations of energy which were generally designated as "electric" or more precisely "electromagnetic" are energy manifestations of the same nature as those of radiant heat and light. Therefore the phenomena of light and heat and others besides these, may be called electrical phenomena. Thus electrical science has become the mother science of all and its study has become all important. The *day* when we shall know exactly what "electricity" is, will chronicle an event probably greater, more important than any other recorded in the history of the human race. The time will come when the comfort, the very existence, perhaps, or man will depend upon that wonderful agent. For our existence and comfort we require heat, light and mechanical power. How do we now get all these? We get them from fuel, we get them by consuming material. What will man do when the forests disappear, when the coal fields are exhausted? Only one thing according to our present knowledge will remain; that is, to transmit power at great distances. Men will go to the waterfalls, to the tides, which are the stores of an infinitesimal part of Nature's immeasurable energy. There will they harness the energy and transmit the same to their settlements, to warm their homes by, to give them light, and to keep their obedient slaves, the machines, toiling. But how will they transmit this energy if not by electricity? judge then, if the comfort, nay, the very existence, of man will not depend on electricity. I am aware that this view is not that of a practical engineer, but neither is it that of an illusionist, for it is certain, that power transmission, which at present is merely a stimulus to enterprise, will some day be a dire necessity.

It is more important for the student, who takes up the study of light phenomena, to make himself thoroughly acquainted with certain modern views, than to peruse entire books on the subject of light itself, as disconnected from these views. Were I therefore to make these demonstrations before students seeking information — and for the sake of the few of those who may be present, give me leave to so assume — it would be my principal endeavor to impress these views upon their minds in this series of experiments.

It might be sufficient for this purpose to perform a simple and well-known experiment. I might take a familiar appliance, a Leyden jar, charge it from a frictional machine, and then discharge it. In explaining to you its permanent state when charged, and its transitory condition when discharging, calling your attention to the forces which enter into play and to the various phenomena they produce, and pointing out the relation of the forces and phenomena, I might fully succccd in illustrating that modern idea. No doubt, to the thinker, this simple experiment would appeal as much as the most magnificent display. But this is to be an experimental demonstration, and one which should possess, besides instructive, also entertaining features and as such, a simple experiment, such as the one cited, would not go very far towards the attainment of the lecturer's aim. I must therefore choose another way of illustrating, more spectacular certainly, but perhaps also more instructive. Instead of the frictional machine and Leyden jar, I shall avail myself in these experiments, ,of an induction coil of peculiar properties, which was described in detail by me in a lecture before the London Institution of Electrical Engineers, in Feb., 1892. This induction coil is capable of yielding currents of enormous potential differences, alternating with extreme rapidity. With this apparatus I shall endeavor to show you three distinct classes of effects, or phenomena, and it is my desire that each experiment, while serving for the purposes of illustration, should at the. same time teach us some novel truth, or show us some novel aspect of this fascinating science. But before doing this, it seems proper and useful to dwell upon the apparatus employed, and method of obtaining the high potentials and high-frequency currents which are made use of in these experiments.

ON THE APPARATUS AND METHOD OF CONVERSION.

These high-frequency currents are obtained in a peculiar manner. The method employed was advanced by me about two years ago in an experimental lecture before the American Institute of Electrical Engineers.

Fig. 1.

A number of ways, as practiced in the laboratory, of obtaining these currents either from continuous or low frequency alternating currents, is diagrammatically indicated in Fig. 1, which will be later described in detail. The general plan is to charge condensers, from a direct or alternate-current source, preferably of high-tension, and to discharge them disruptively while observing well-known conditions necessary to maintain the oscillations of the current. In view of the general interest taken in high-frequency currents and effects producible by them, it seems to me advisable to dwell at some

237

length upon this method of conversion. In order to give you a clear idea of the action, I will suppose that a continuous-current generator is employed, which is often very convenient. It is desirable that the generator should possess such high tension as to be able to break through a small air space. If this is not the case, then auxiliary means have to be resorted to, some of which will be indicated subsequently. When the condensers are charged to a certain potential, the air, or insulating space, gives way and a disruptive discharge: occurs. There is then a sudden rush of current and generally a large portion of accumulated electrical energy spends itself. The condensers are thereupon quickly charged and the same process is repeated in more or less rapid succession. To produce such sudden rushes of current it is necessary to observe certain conditions. If the rate at which the condensers are discharged is the same as that at which they are charged, then, clearly, in the assumed case the condensers do not come into play. If the rate of discharge be smaller than the rate of charging, then, again, the condensers cannot play an important part. But if, on the contrary, the rate of discharging is greater than that of charging, then a succession of rushes of current is obtained. It is evident that, if the rate at which the energy is dissipated by the discharge is very much greater than the rate of supply to the condensers, the sudden rushes will be comparatively few, with long-time intervals between. This always occurs when a condenser of considerable capacity is charged by means of a comparatively small machine. If the rates of supply and dissipation are not widely different, then the rushes of current will be in quicker succession, and this the more, the more nearly equal both the rates are, until limitations incident to each case and depending upon a number of causes are reached. Thus we are able to obtain from a continuous-current generator as rapid a succession of discharges as we like. Of course, the higher the tension of the generator, the smaller need be the capacity of the condensers, and for this reason, principally, it is of advantage to employ a generator of very high tension. Besides, such a generator permits the attaining of greater rates of vibration.

The rushes of current may be of the same direction under the conditions before assumed, but most generally there is an oscillation superimposed upon the fundamental vibration of the current. When the conditions are so determined that there are no oscillations, the current impulses are unidirectional and thus a means is provided of transforming a continuous current of high tension, into a direct current of lower tension, which I think may find employment in the arts.

This method of conversion is exceedingly interesting and I was much impressed by its beauty when I first conceived it. It is ideal in certain respects. It involves the employment of no mechanical devices of any kind, and it allows of obtaining currents of any desired frequency from an ordinary circuit, direct or alternating. The frequency of the fundamental discharges depending on the relative rates of supply and dissipation can be readily varied within wide limits, by simple adjustments of these quantities, and the frequency of the superimposed vibration by the determination of the capacity, self-induction and resistance of the circuit. The potential of the currents, again, may be raised as high as *any* insulation is capable of withstanding safely by combining capacity and self-induction or by induction in a secondary, which need have but comparatively few turns.

As the conditions are often such that the intermittence or oscillation does not readily establish itself, especially when a direct current source is employed, it is of advantage to associate an interrupter with the arc, as I have, some time ago, indicated the use of an air-blast or magnet, or other such device readily at hind. The magnet is employed with special advantage in the conversion of direct currents, as it is then very effective. If the primary source is an alternate current generator. it is desirable, as I have stated on another occasion, that the frequency should be low, and that tile current forming the arc be large, in order to render the magnet more effective.

A form of such discharger with a magnet which has been found convenient, and adopted after some trials, in the conversion of direct currents particularly, is illustrated in Fig. 2. N S are the pole pieces of a very strong magnet which is excited by a coil c. The pole pieces are slotted for adjustment and can be fastened in any position by screws s sl. The discharge rods d d1, thinned down on the ends in order to allow a closer approach of the magnetic pole pieces, pass through the columns of brass b b1 and are fastened in position by screws s_2 s_2. Springs *r*

r_1 and collars c c_1 are slipped on the rods, the latter serving to set the points of the rods at a certain suitable distance by means of screws s_3 $s_3$and the former to draw the points apart.

Fig. 2.

When it is desired to start the arc, one of the large rubber handles h h_1 is tapped quickly with the hand, whereby the points of the rods are brought in contact but are instantly separated by the springs r r_1. Such an arrangement has been found to be often necessary, namely in cases when the E. M. F. was not large enough to cause the discharge to break through the gap, and also when it was desirable to avoid short circuiting of the generator by the metallic contact of the rods. The rapidity of the interruptions of the current with a magnet depends on the intensity of the magnetic field and on the potential difference at the end of the arc. The interruptions are generally in such quick succession as to produce a musical sound.

Years ago it was observed that when a powerful induction coil is discharged between the poles of a strong magnet, the discharge produces a loud noise not unlike a small pistol shot. It was vaguely stated that the spark was intensified by the presence of the magnetic field. It is now clear that the discharge current, flowing for some time, was interrupted a great number of times by the magnet, thus producing the sound. The phenomenon is especially marked when the field circuit of a large magnet or dynamo is broken in a powerful magnetic field.

When the current through the gap is comparatively large, it is of advantage to slip on the points of the discharge rods pieces of *very* hard carbon and let the arc play between the carbon pieces. This preserves the rods, and besides has the advantage of keeping the air space hotter, as the heat is not conducted away as quickly through the carbons, and the result is that a smaller E. M. F. in the arc gap is required to maintain a succession of discharges.

Another form of discharger, which may be employed with advantage in some cases, is illustrated in Fig. 3. In this form the discharge rods d d_1 pass through perforations in a wooden box B, which is thickly coated with mica on the inside, as indicated by the heavy lines. The perforations are provided with mica tubes m m_1 of some thickness, which are preferably not in contact with the rods d d_1. The box has a cover c which is a little larger and descends on the outside of the box. The spark gap is warmed by a small lamp l contained in the box. A plate p above the lamp allows the draught to pass only through the chimney a of the lamp, the air entering through holes o o in or near the bottom of the box and following the path indicated by the arrows. When the discharger is in operation, the door of the box is closed so that the light of the arc is not visible outside. It is desirable to exclude the light as perfectly as possible, as it interferes with some experiments. This form of discharger is simple and very effective when properly manipulated. The air being warmed to a certain temperature, has its insulating power impaired; it becomes dielectrically weak, as it were, and the consequence is that the arc can be established at much greater distance. The arc should, of course, be sufficiently insulating to allow the discharge to pass through the gap *disruptively.* The arc formed under such conditions, when long, may be made extremely sensitive, and the weal: draught through the lamp chimney a is quite sufficient to produce rapid interruptions. The adjustment is made by regulating the temperature and velocity of the draught. Instead of using the lamp, it answers the purpose to provide for a draught of warm air in other ways. A very simple way which has been practiced is to enclose the arc in a long vertical tube, with plates on the top and bottom for regulating the temperature and velocity of the air current. Some provision had to be made for deadening the sound.

The air may be rendered dielectrically weak also by rarefaction. Dischargers of this kind have likewise been used by me in connection with a magnet. A large tube is for this purpose provided with heavy electrodes of carbon or metal, between which the discharge is made to pass, the tube being placed in a powerful magnetic field.

Fig. 3.

The exhaustion of the tube is carried to a point at which the discharge breaks through easily, but the pressure should be more than 75 millimetres, at which the ordinary thread discharge occurs. In another form of discharger, combining the features before mentioned, the discharge was made to pass between two adjustable magnetic pole pieces, the space between them being kept at an elevated temperature.

It should be remarked here that when such, or interrupting devices of any kind, are used and the currents are passed through the primary of a disruptive discharge coil, it is not, as a rule, of advantage to produce a number of interruptions of the current per second greater than the natural frequency of vibration of the dynamo supply circuit, which is ordinarily small. It should also be pointed out here, that while the devices mentioned in connection with the disruptive discharge are advantageous under certain conditions, they may be sometimes a source of trouble, as they produce intermittences and other irregularities in the vibration which it would be very desirable to overcome.

There is, I regret to say, in this beautiful method of conversion a defect, which fortunately is not vital, .and which I have been gradually overcoming. I will best call attention to this defect and indicate a fruitful line of work, by comparing the electrical process with its mechanical analogue. The process may be illustrated in this manner. Imagine a tank; with a wide opening at the bottom, which is kept closed by spring pressure, but so that it snaps off *sudden/y* when the liquid in the tank has reached a certain height. Let the fluid be supplied to the tank by means of a pipe feeding at a certain rate. When the critical height of the liquid is reached, the spring gives way and the bottom of the tank drops out. Instantly the liquid falls through the wide opening, and the spring, reasserting itself, closes the bottom again. The tank is now filled, and after a certain time interval the same process is repeated. It is clear, that if the pipe feeds the fluid quicker than the bottom outlet is capable of letting it pass through, the bottom will remain off and the tank; will still overflow. If the rates of supply are exactly eual, then the bottom lid will remain partially open and no vibration of the same and of the liquid column will generally occur, though it might, if started by some means. But if the inlet pipe does not feed the fluid fast enough for the outlet, then there will be always vibration. Again, in such case, each time the bottom flaps up or down, the spring and the liquid column, if the pliability of the spring and the inertia of the moving parts are properly chosen, will perform independent vibrations. In this analogue the fluid may be likened to electricity or electrical energy, the tank to the condenser, the spring to the dielectric, and the pipe to the conductor through which electricity is supplied to the condenser. To make this analogy quite complete it is necessary to make the assumption, that the bottom, each time it gives way, is knocked violently against a non-elastic stop, this irnpact involving some loss of energy; and that, besides, some dissipation of energy results due to frictional losses. In the preceding analogue the liquid is supposed to be under a steady pressure. If the presence of the fluid be assumed to vary rhythmically, this may be taken as corresponding to the case of an alternating current. The process is then not quite as simple to consider, but the action is the same in principle.

It is desirable, in order to maintain the vibration economically, to reduce the impact and frictional losses as much as possible. As regards the latter, which in the electrical analogue correspond to the losses due to the resistance of the circuits, it is impossible to obviate them entirely, but they can be reduced to a minimum by a proper selection of the dimensions of the circuits and by the employment of thin conductors in the form of strands. But the loss of energy caused by the first breaking through of the dielectric — which in the above example corresponds to the violent knock of the bottom against the inelastic stop — would be more important to overcome. At the moment of the breaking through, the air space has a very high resistance, which is probably reduced to a very small value when the current has reached some strength, and the space is brought to a high temperature. It would materially diminish the loss of energy if the space were always kept at an extremely high temperature, but then there would be no disruptive break. By warming the space moderately by means of a lamp or otherwise, the economy as far as the arc is concerned is sensibly increased. But the magnet or other interrupting device does not diminish the loss in the arc. Likewise, a jet of air only facilitates the carrying off of the energy. Air, or a gas in ,eneral, behaves curiously in this respect. When two bodies charged to a very high potential, discharge disruptively through an air space, any amount of energy may be carried off by the air. This energy is evidently dissipated by bodily carriers, in impact and collisional losses of the molecules. The exchange of the molecules in the space occurs with inconceivable rapidity. A powerful discharge taking place between two electrodes, they may remain entirely cool, and yet the loss in the air may represent any amount of energy. It is perfectly practicable, with very great potential differences in the gap, to dissipate several horse-power in the arc of the discharge without even noticing a small increase in the temperature of the electrodes. All the frictional losses occur then practically in the air. If the exchange of the air molecules is prevented, as by enclosing the air hermetically, the bas inside of the vessel is brought quickly to a high temperature, even with a very small discharge. It is difficult to estimate how much of the energy is lost in sound waves, audible or not, in a powerful discharge. When the currents through the gap are large, the electrodes may become rapidly heated, but this is not a reliable measure of the energy wasted in the arc, as the loss through the yap itself may be comparatively small. The air or a gas in general is at ordinary pressure at least, clearly not the best medium through which a disruptive discharge should occur. Air or other gas under great pressure is of curse a much more suitable medium for the discharge gap. I have carried on long-continued experiments in this direction, unfortunately less practicable on account of the difficulties and expense in getting air under great pressure. But even if the medium in the discharge space is solid or liquid, still the same losses take place, though they are generally smaller, for just as soon as the arc is established, the solid or liquid is volatilized. Indeed, !here is no body known which would not be disintegrated by the arc, and it is an open question among scientific men, whether an arc discharge could occur at all in the air itself without the particles of the electrodes being torn off. When the current through the gap is very small and the arc very long, I believe that a relatively considerable amount of heat is taken up in the disintegration of the electrodes, which partially on this account may remain quite cold.

The ideal medium for a discharge gap should only *crack.* and the ideal electrode should be of some material which cannot be disintegrated. With small currents through the gap it is best to employ aluminum, but not when the currents are large. The disruptive break in the air, or more or less in any ordinary medium, is not of the nature of a crack, but it is rather comparable to the piercing of innumerable bullets through a mass offering great frictional resistances to the motion of the bullets, this involving considerable loss of energy. A medium which would merely crack when strained electrostatically — and this possibly might be the case with a perfect vacuum, that is, pure ether — would involve a very small loss in the gap, so small as to be entirely negligible, at least theoretically, because a crack may be produced by an infinitely small displacement. In exhausting an oblong bulb provided with two aluminum terminals, with the greatest care, I have succeeded in producing such a vacuum that the secondary discharge of a disruptive discharge coil would break disruptively through the bulb in the form of fine spark streams. The curious point was that the discharge would completely ignore the terminals and start far behind the two aluminum plates which served as electrodes. This extraordinary high vacuum could only be maintained for a very short while. To return to the ideal medium; think, for the sake of illustration, of a piece of glass or similar body clamped in a vice, and the latter tightened more and more. At a certain point a minute increase of the pressure will cause the ;glass to crack. The loss of energy involved in splitting the glass may be practically nothing, for though the force is great, the displacement need be but extremely small. Now imagine that the glass would possess the property of closing again perfectly the crack upon a minute diminution of the pressure. This is the way the dielectric in the discharge space should behave. But inasmuch as there would be always some loss in the gap, the medium, which should be continuous should exchange through the gap at a rapid rate. In the preceding example, the glass being perfectly closed, it would mean that the dielectric in the discharge space possesses a great insulating power; the glass being cracked, it would signify that the medium in the space is a good conductor. The dielectric should vary enormously in resistance by minute variations of the E. M. F. across the discharge space. This condition is attained, but in an extremely imperfect manner, by warming the air space to a. certain critical temperature, dependent on the E. M. F. across the yap, or by otherwise impairing the insulating power of the air. But as a matter of fact the air does never break down *disruptively,* if this term be rigorously interpreted, for before the sudden rush of the current occurs, there is always a weak current preceding it, which rises first gradually and then with comparative suddenness. That is the reason why the rate of change is very much greater when glass, for instance, is broken through, than when the break takes place through an air space of equivalent dielectric strength. As a medium for the discharge space, a solid, or even a liquid, would be preferable therefore. It is somewhat difficult to conceive of a solid body which would possess the property of closing instantly after it has been cracked. But a liquid, especially under great pressure, behaves practically like a solid, while it possesses the property of closing the crack. Hence it was thought that a liquid insulator might be more suitable as a dielectric than air. Following out this idea, a number of different forms of dischargers in which a variety of such insulators, sometimes under great pressure, were employed, have been experimented upon. It is thought sufficient to dwell in a few words upon one of the forms experimented upon. One of these dischargers is illustrated in Figs. 4a and 4b.

A hollow metal pulley P (Fig. 4a), was fastened upon an arbor a, which by suitable means was rotated at a considerable speed. On the inside of the pulley, but disconnected from the same, was supported a thin disc h (which is shown thick for the sake of clearness), of hard rubber in which there were embedded two metal segments s s with metallic extensions *e e* into which were screwed conducting terminals *t t* covered with thick tubes of hard rubber *t t*.

Fig. 4a.   Fig. 4b.

The rubber disc b with its metallic segments s s, was finished in a lathe, and its entire surface highly polished so as to offer the smallest possible frictional resistance to the motion through a fluid. In the hollow of the pulley an insulating liquid such as a thin oil was poured so as to reach very nearly to the opening left in the flange f, which was screwed tightly on the front side of the pulley. The terminals t t, were connected to the opposite coatings of a battery of condensers so that the discharge occurred through the liquid. When the pulley was rotated, the liquid was forced against the rim of the pulley and considerable fluid pressure resulted. In this simple way the discharge gap was filled with a medium which behaved practically like a solid, which possessed the duality of closing instantly upon the occurrence of the break, and which moreover was circulating through the gap at a rapid rate. Very powerful effects were produced by discharges of this kind with liquid interrupters, of which a number of different forms were made. It was found that, as expected, a longer spark for a given length of wire was obtainable in this way than by using air as an interrupting device. Generally the speed, and therefore also the fluid pressure, was limited by reason of the fluid friction, in the form of discharger described, but the practically obtainable speed was more than sufficient to produce a number of breaks suitable for the circuits ordinarily used. In such instances the metal pulley P was provided with a few projections inwardly, and a definite number of breaks was then produced which could be computed from the speed of rotation of the pulley. Experiments were also carried on with liquids of different insulating power with the view of reducing the loss in the arc. When an insulating liquid is moderately warmed, the loss in the arc is diminished.

A point of some importance was noted in experiments with various discharges of this kind. It was found, for instance, that whereas the conditions maintained in these forms were favorable for the production of a great spark length, the current so obtained was not best suited to the production of light effects. Experience undoubtedly has shown, that for such purposes a harmonic rise and fall of the potential is preferable. Be it that a solid is rendered incandescent, or phosphorescent, or be it that energy is transmitted by condenser coating through the glass, it is quite certain that a harmonically rising and falling potential produces less destructive action, and that the vacuum is more permanently maintained. This would be easily explained if it were ascertained that the process going on in an exhausted vessel is of an electrolytic nature. In the diagrammatical sketch, Fig. 1, which has been already referred to, the cases which are most likely to be met with in practice are illustrated. One has at his disposal either direct or alternating currents from a supply station. It is convenient for an experimenter in an isolated laboratory to employ a machine G, such as illustrated, capable of giving both kinds of currents. In such case it is also preferable to use a machine with multiple circuits, as in many experiments it is useful and convenient to have at one's disposal currents of different phases. In the sketch, D represents the direct and A the alternating circuit. In each of these, three branch circuits are shown, all of which are provided with double line switches s s s s s s. Consider first the direct current conversion; la represents the simplest case. If the E. M. F. of the generator is sufficient to break through a small air space, at least when the latter is warmed or otherwise rendered poorly insulating, there is no difficulty in maintaining a vibration with fair economy by judicious adjustment of the capacity, self-induction and resistance of the circuit L. containing the devices *l l m.* The magnet N, S, can be in this case advantageously combined with the air space, The discharger d d with the magnet may be placed either way, as indicated by the full or by the dotted lines. The circuit la with the connections and devices is supposed to possess dimensions such as are suitable for the maintenance of a vibration. But usually the E. M. F. on the circuit or branch la will be something like a 100 volts or so, and in this case it is not sufficient to break through the gap. Many different means may be used to remedy this by raising the E. M. F. across the gap. The simplest is probably to insert a large self-induction coil in series with the circuit L. When the arc is established, as by the discharger illustrated in Fig. 2, the magnet blows the arc out the instant it is formed. Now the extra current of the break, being of high E. M. F., breaks through the gap, and a path of low resistance for the dynamo current being again provided, there is a sudden rush of current from the dynamo upon the weakening or subsidence of the extra current. This process is repeated in rapid succession, and in this manner I have maintained oscillation with as low as 50 volts, or even less, across the gap. But conversior on this plan is not to be recommended on account of the too heavy currents through the gap and consequent heating of the electrodes; besides, the frequencies obtained in this way are low, owing to the high self-induction necessarily associated with the circuit. It is very desirable to have the E. M. F. as high as possible, first, in order to increase the economy of the conversion, and secondly, to obtain high frequencies. The difference of potential in this electric oscillation is, of course, the equivalent of the stretching force in the mechanical vibration of the spring. To obtain very rapid vibration in a circuit of some inertia, a great stretching force or difference of potential is necessary. Incidentally, when the E. M. F. is very great, the condenser which is usually employed in connection with the circuit need but have a small capacity, and many other advantages are gained. With a view of raising the E. M. F. to a many times greater value than obtainable from ordinary distribution circuits, a rotating transformer g is used, as indicated at I 1a. Fig. 1, or else a separate high potential machine is driven by means of a motor operated from the generator G. The latter plan is in fact preferable, as changes are easier made. The connections from the high tension winding are quite similar to those in branch la with the exception that a condenser C, which should be adjustable, is connected to the high tension circuit. Usually, also, an adjustable self-induction coil in series with the circuit has been employed in these experiments. When the tension of the currents is very high, the magnet ordinarily used in connection with the discharger is of comparatively small value, as it is quite easy to adjust the dimensions of the circuit so that oscillation is maintained. The employment of a steady E. M. F. in the high frequency conversion affords some advantages over the employment of alternatin E. M. F., as the adjustments are much simpler and the action can be easier controlled. But unfortunately one is limited by the obtainable potential difference. The winding also breaks down easily in consequence of the sparks which form between the sections of the armature or commutator when a vigorous oscillation takes place. Besides, these transformers are expensive to build. It has been found by experience that it is best to follow the plan illustrated at Met, In this arrangement a rotating transformer ,g, is employed to convert the low tension direct currents into low frequency alternating currents, preferably also of small tension. The tension of the currents is then raised in a stationary transformer T. The secondary s of this transformer is connected to an adjustable condenser C which discharges through the gap or discharger d d, placed in either of the ways indicated, through the primary P of a disruptive discharge coil, the high frequency current being obtained from the secondary s of this coil, as described on previous occasions. This will undoubtedly be found the cheapest and most convenient way of converting direct currents.

The three branches of the circuit A represent the usual cases met in practice when alternating currents are converted. In Fig. I b a condenser C, generally of large capacity, is connected to the circuit L containing the devices *l l, m m* The devices m m are supposed to be of high self-induction so as to bring the frequency of the circuit more or less to that of the dynamo. In this instance the discharger *d d* should best have a number of makes and breaks per second equal to twice the frequency of the dynamo. If not so, then it should have at least a number equal to a multiple or even fraction of the dynamo frequency. It should be observed, referring to I b, that the conversion to a high potential is also effected when the discharger *d d,* which is shown in the sketch, is omitted. But the effects which are produced by currents which rise instantly to high values, as in a disruptive discharge, are entirely different from those produced by dynamo currents which rise and fall harmonically. So, for instance, there might be in a given case a number of makes and breaks at *d d* equal to just twice the frequency of the dynamo, or in other words, there may be the same number of fundamental oscillations as would be produced without the discharge gap, and there might even not be any quicker superimposed vibration; yet the differences of potential at the various points of the circuit, the impedance and other phenomena, dependent upon the rate of change, will bear no similarity in the two cases. Thus, when working with currents discharging disruptively, the element chiefly to be considered is not the frequency, as a student might be apt to believe, but the rate of change per unit of time. With low frequencies in a certain measure the same effects may be obtained as with high frequencies, provided the rate of change is sufficiently great. So if a low frequency current is raised to a potential of, say, 75,000 volts, and the high tension current passed through a series of high resistance lamp filaments, the importance of the rarefied gas surrounding the filament is clearly noted, as will be seen later; or, if a low frequency current of several thousand amperes is passed through a metal bar, striking phenomena of impedance are observed, just as with currents of high frequencies. But it is, of course, evident that with low frequency currents it is impossible to obtain such rates of change per unit of time as with high frequencies, hence the effects produced by the latter are much more prominent. It is deemed advisable to make the preceding remarks, inasmuch as many more recently described effects have been unwittingly identified with high frequencies. Frequency alone in reality does not mean anything, except when an undisturbed harmonic oscillation is considered.

In the branch III b a similar disposition to that in Ib is illustrated, with the difference that the currents discharging through the gap *d d* are used to induce currents in the secondary s of a transformer T. In such case tale secondary should be provided with an adjustable condenser for the purpose of tuning it to the primary.

II b illustrates a plan of alternate current high frequency conversion which is most frequently used and which is found to be most convenient. This plan has been dwelt upon in detail on previous occasions and need not be described here.

Some of these results were obtaincd by the use of a high frequency alternator. A description of such machines will be found in my original paper before the American Institute of Electrical Engineers, and in periodicals of that period, notably in The Electrical Engineer of March 18, 1891.

I will now proceed with the experiments.

ON PHENOMENA PRODUCED BY ELECTROSTATIC FORCE.

The first class of effects I intend to show you are effects produced by electrostatic force. It is the force which governs the motion of the atoms, which causes them to collide and develop the life-sustaining energy of heat and light, and which causes them to aggregate in an infinite variety of ways, according to Nature's fanciful designs, and to form all these wondrous structures we perceive around us; it is, in fact, if our present views be true, the most important force for us to consider in. Nature. As the term *electrostatic* might imply a steady electric condition, it should be remarked, that in these experiments the force is not constant, but varies at a rate which may be considered moderate, about one million times a second, or thereabouts. This enables me to produce many effects which are not producible with an unvarying force.

When two conducting bodies are insulated and electrified, we say that an electrostatic force is acting between them. This force manifests itself in attractions, repulsions and stresses in the bodies and space or medium without. So great may be the strain exerted in the air, or whatever separates the two conducting bodies, that it may break down, and we observe sparks or bundles of light or streamers, as they are called. These streamers form abundantly when the force through the air is rapidly varying. I will illustrate this action of electrostatic force in a novel experiment in which I will employ the induction coil before referred to. The coil is contained in a trough filled with oil, and placed under the table. The two ends of .the secondary wire pass through the two thick columns of hard rubber which protrude to solve height above the table. It is necessary to insulate the ends or terminals of the secondary heavily with hard rubber, because even dry wood is by far ton poor an insulator for these currents of enormous potential differences. On one of the terminals of the coil, I ,have placed a large sphere of sheet brass, which is connected to a larger insulated brass plate, in order to enable me to perform the experiments under conditions, which, as you will see, are more suitable for this experiment. I now set the coil to work and approach the free terminal with a metallic object held in my hand, this simply to avoid burns As I approach the metallic object to a distance of eight or tell inches, a torrent of furious sparks breaks forth from the end of the secondary wire, which passes through the rubber column. The sparks cease when the metal in my hand touches the wire. My arm is now traversed *by* a powerful electric current, vibrating at about the rate of one million times a second. All around me the electrostatic force makes itself felt, and the air molecules and particles of dust flying about are acted upon and are hammering violently against my body. So great is this agitation of the particles, that when the lights are turned out you may see streams of feeble light appear on some parts of my body. When such a streamer breaks out on any part of the body, it produces a sensation like the pricking of a needle. Were the potentials sufficiently high and the frequency of the vibration rather low, the skin would probably be ruptured under the tremendous strain, and the blood would rush out with great force in the form of fine spray or jet so thin as to be invisible, just as oil will when placed on the positive terminal of a Holtz machine. The breaking through of the skin though it may seem impossible at first, would perhaps occur, by reason of the tissues finder the skin being incomparably better conducting. This, at least, appears plausible, judging from some observations.

Fig. 5.

I can make these streams of light visible to all, by touching with the metallic object one of the terminals as before, and aproaching my free hand to the brass sphere, which is connected to the second terminal of the coil. As the hand is approached, the air between it and the sphere, or in the immediate neighborhood, is more violently agitated, and you see streams of light now break forth from my finger tips and from the whole hand (Fig. 5). Were I to approach the hand closer, powerful sparks would jump from the brass sphere to my hand, which might be injurious. The streamers offer no particular inconvenience, except that in the ends of the finger tips a burning sensation is felt. They should not be confounded with those produced by an influence machine, because in many respects they behave differently. I have attached the brass sphere and plate to one of the terminals in order to prevent the formation of visible streamers

262

on that terminal, also in order to prevent sparks from jumping at a considerable distance. Besides, the attachment is favorable for the working of the coil.

The streams of light which you have observed issuing from my hand are due to a potential of about 200,000 volts, alternating in rather irregular intervals, sometimes like a million times a second. A vibration of the same amplitude, but four times as fast, to maintain which over 3,000,000 volts would be required, would be mare than sufficient to envelop my body in a complete sheet of flame. But this flame would not burn me up; quite contrarily, the probability is ,that I would not be injured in the least. Yet a hundredth part of that energy, otherwise directed; would be amply sufficient to kill a person.

The amount of energy which may thus be passed into the body of a person depends on the frequency and potential of the currents, and by making both of these very great, a vast amount of energy may be passed into the body without causing any discomfort, except perhaps, in the arm, which is traversed by a true conduction current. The reason why no pain in the body is felt, and no injurious effect noted, is that everywhere, if a current be imagined to flow through the body, the direction of its flow would be at right angles to the surface; hence the body of the experimenter offers an enormous section to the current, and the density is very small, with the exception of the arm, perhaps, where the density may be considerable. Lout if only a small fraction of that energy would be applied in such a way that a current would traverse the body in the same manner as a low frequency current, a shock would be received which might be fatal. A direct or low frequency alternating current is fatal, I think, principally because its distribution through the body is not uniform, as it must divide itself in minute streamlets of great density, whereby some organs are vitally injured. That such a process occurs I have not the least doubt, though no evidence might apparently exist, or be found upon examination. The surest to injure and destroy life, is a continuous current, but the most painful is an alternating current of very low frequency. The expression of these views, which are the result of long continued experiment and observation, both with steady and varying currents, is elicited by the interest which is at present taken in this subject, and by the manifestly erroneous ideas which are daily propounded in journals on this subject.

I may illustrate an effect of the electrostatic force by another striking experiment, but before, I must call your attention to one or two facts. I have said before, that whet; the medium between two oppositely electrified bodies is strained beyond a certain limit it gives way and, stated in popular language, the opposite electric charges unite and neutralize each other. This breaking down of the medium occurs principally when the force acting between the bodies is steady, or varies at a moderate rate. Were the variation sufficiently rapid, such a destructive break would not occur, no matter how great the force, for all the energy would be spent in radiation, convection and mechanical and chemical action. Thus the *spark* length, or greatest distance which *a spark* will jump between the electrified bodies is the smaller, the greater the variation or time rate of change. But this rule may be taken to be true only in a general way, when comparing rates which are widely different.

Fig. 6a.   Fig. 6b.

I will show you by an experiment the difference in the effect produced by a rapidly varying and a steady or moderately varying force. I have here two large circular brass plates *p p* (Fig. 6a and Fig. 6b), supported on movable insulating stands on the table, connected to the ends of the secondary of a coil similar to the one used before. I place the plates ten or twelve inches apart and set the coil to work. You see the whole space between the plates, nearly two cubic feet, filled with uniform light, Fig. 6a. This light is due to the streamers you have seen in the first experiment, which are now much more intense. I have already pointed out the importance of these streamers in commercial apparatus and their still greater importance in some purely scientific

265

investigations, Often they are to weak to be visible, but they always exist, consuming energy and modifying the action of .the apparatus. When intense, as they are at present, they produce ozone in great quantity, and also, as Professor Crookes has pointed out, nitrous acid. So quick is the chemical action that if a coil, such as this one, is worked for a very long time it will make the atmosphere of a small room unbearable, for the eyes and throat are attacked. But when moderately produced, the streamers refresh the atmosphere wonderfully, like a thunder-storm, and exercises unquestionably a beneficial effect.

In this experiment the force acting between the plates changes in intensity and direction at a very rapid rate. I will now make the rate of change per unit time much smaller. This I effect by rendering the discharges through the primary of the induction coil less frequent, and also by diminishing the rapidity of the vibration in the secondary. The former result is conveniently secured by lowering the E. M. F. over the air gap in the primary circuit, the latter by approaching the two brass plates to a distance of about three or four inches. When the coil is set to work, you see no streamers or light between the plates, yet the medium between them is under a tremendous strain. I still further augment the strain by raising the E. M F. in the primary circuit, and soon you see the air give away and the hall is illuminated by a shower of brilliant and noisy sparks, Fig. 6b. These sparks could be produced also with unvarying force; they have been for many years a familiar phenomenon, though they were usually obtained from an entirely different apparatus. In describing these two phenomena so radically different in appearance, I have advisedly spoken of a "force" acting between the plates. It would be in accordance with the accepted views to say, that there was an "alternating E. M. F.", acting between the plates. This term is quite proper and applicable in all cases where there is evidence of at least a possibility of an essential inter-dependence of the electric state of the plates, or electric action in their neighborhood. But if the plates were removed to an infinite distance, or if at a finite distance, there is no probability or necessity whatever for such dependence. I prefer to use the term "electrostatic force," and to say that such a force is acting around *each* plate or electrified insulated body in general. There is an inconvenience in usin this expression as the term incidentally, means a steady electric condition; but a proper nomenclature will eventually settle this difficulty.

Fig. 7.

I now return to the experiment to which I have already alluded, and with which I desire to illustrate a striking effect produced by a rapidly varying electrostatic force. I attach to the end of the wire, l (Fig. 7), which is in connection with one of the terminals of the secondary of the induction coil, an exhausted bulb b. This bulb contains a thin carbon filament f, which is fastened to a platinum wire w, sealed in the glass and leading outside of the bulb, where it connects to the wire l. The bulb may be exhausted to any degree attainable with ordinary apparatus. Just a moment before, you have witnessed the breaking down of the air between the charged brass plates. You know that a plate of glass, or any other insulating material, would break down in like manner. Had I therefore a metallic coating attached to the outside of the bulb, or placed near the same, and were this coating connected to the other terminal of the coil, would be prepared to see the glass give way if the strain were sufficiently increased. Even were the coating not connected to the other terminal, but to an insulated plate, still, if you have followed recent developments, you would naturally expect a rapture of the glass.

But it will certainly surprise you to note that under the action of the varying electrostatic force, the glass gives way when all other bodies are removed from the bulb. In fact, all the surrounding bodies we perceive might be removed to an infinite distance without affecting the result in the slightest. When the coil is set to work, the glass is invariably broken through at the seal, or other narrow channel, and the vacuum is quickly impaired. Such a damaging break would not occur with a steady force, even if the same were many times greater. The break is due to the agitation of the molecules of the gas within the bulb, and outside of the same. This agitation, which is generally most violent in the narrow pointed channel near the seal, causes a heating and rupture of the glass. This rupture would, however, not occur, not even with a varying force, if the medium filling the inside of the bulb, and that surrounding it, were perfectly homogeneous. The break occurs much quicker if the top of the bulb is drawn out into a fine fibre. In bulbs used with these coils such narrow, pointed channels must therefore be avoided.

Fig. 8a.   Fig. 8b.

When a conducting body is immersed in air, or similar insulating medium, consisting of, or containing, small freely movable particles capable of being electrified, and when the electrification of the body is made to undergo a very rapid change — which is equivalent to saying that the electrostatic force acting around the body is varying in intensity, — the small particles are attracted and repelled, and their violent impacts against the body may cause a mechanical motion of the latter. Phenomena of this kind are noteworthy, inasmuch as they have not been observed before with apparatus such as has been commonly in use. If a very light conducting sphere be suspended oil an exceedingly fine wire, and charged to a steady potential, however high, the sphere will remain at rest. Even if the potential would be rapidly varying, provided that the small particles of matter, molecules or atoms, are evenly distributed, no motion of the sphere should result. But if one side of the conducting sphere is covered with a thick insulating *layer,* the impacts of the particles will cause the sphere to move about, generally in irregular curves, Fig. 8a. In like manner, as I have shown on a previous occasion, a fan of sheet metal, Fig. 8b, covered partially with insulating material as indicated, and placed upon the terminal of the coil so as to turn freely on it, is spun around.

Fig. 9.

All these phenomena you have witnessed and others which will be shown later, are due to the presence of a medium like air, and would not occur in a continuous medium. The action of the air may be illustrated still better by the following experiment. I take a glass tube *t,* Fig. 9, of about an inch in diameter, which has a platinum wire w sealed in the lower end, and to which is attached a thin lamp filament *f.* I connect the wire with the terminal of the coil and set the coil to work. The platinum wire is now electrified positively and negatively in rapid succession and the wire and air inside of the tube is rapidly heated by the impacts of the particles, which may be so violent as to render the filament incandescent. But if I pour oil in the tube, just as soon as the wire is covered with the oil, all action apparently ceases and there is no marked evidence of heating. The reason of this is that the oil is a practically continuous medium. The displacements in such a continuous medium are, with these frequencies, to all appearance incomparably smaller than in air, hence the work performed in such a medium is insignificant. But oil would behave very differently with frequencies many times as great, for even though the displacements be small, if the frequency were much greater, considerable work might be performed in the oil.

The electrostatic attractions and repulsions between bodies of measurable dimensions are, of all the manifestations of this force, the first so-called *electrical* phenomena noted. But though they have been known to us for many centuries, the precise nature of the mechanism concerned in these actions is still unknown to us, and has not been even quite satisfactorily explained. What kind of mechanism must that be? We cannot help wondering when we observe two magnets attracting and repelling each other with a force of hundreds of pounds with apparently nothing between them. We have in our commercial dynamos magnets capable of sustaining in mid-air tons of weight. But what are even these forces acting between magnets when compared with the tremendous attractions and repulsions produced by electrostatic force, to which there is apparently no limit as to intensity. In lightning discharges bodies are often charged to so high a potential that they are thrown away with inconceivable force and torn asunder or shattered into fragments. Still even such effects cannot compare with the attractions and repulsions which exist between charged molecules or atoms, and which are sufficient to project them with speeds of many kilometres a second, so that under their violent impact bodies are rendered highly incandescent and are volatilized. It is of special interest for the thinker who inquires into the nature of these forces to note that whereas the actions between individual molecules or atoms occur seemingly under any conditions, the attractions and repulsions of bodies of measurable dimensions imply a medium possessing insulating properties. So, if air; either by being rarefied or heated, is rendered more or less conducting, these actions between two electrified bodies practically cease, while the actions between the individual atoms continue to manifest themselves.

Fig. 10.

An experiment may serve as an illustration and as a means of bringing out other features of interest: Some time ago I showed that a lamp filament or wire mounted in **a** bulb and connected to one of the terminals of a high tension secondary coil is se'r spinning, the top of the filament generally describing a circle. This vibration was very energetic when the air in the bulb was at ordinary pressure and became less energetic when the air in the bulb was strongly compressed. It ceased altogether when the air was exhausted so as to become comparatively good conducting. I found at that time that no vibration tool: place when the bulb was very highly exhausted. But I conjectured that the vibration which I ascribed to the electrostatic action between the walls of the bulb and the filament should take place also in a highly exhausted bulb. To test this under conditions which were more favorable, a bulb like the one in Fig. 10; was constructed. It comprised a globe b, in the neck of which was sealed a platinum wire w, carrying a thin lamp filament f. In the lower part of the globe a tube t was sealed so as to surround the filament. The exhaustion was carried as far as it was practicable with the apparatus employed.

This bulb verified my expectation, for the filament was set spinning when the current was turned on, and became incandescent. It also showed another interesting feature, bearing upon the preceding remarks, namely, when the filament had been kept incandescent some time, the narrow tube and the space inside were brought to an elevated temperature, and as the gas in the tube then became conducting, the electrostatic attraction between the glass and the filament became *very* weak or ceased, and the filament came to rest, When it came to rest it would glow far more intensely. This was probably due to its assuming the position in the centre of the tube where the molecular bombardment was most intense, and also partly to the fact that the individual impacts were more violent and that no part of the supplied energy was converted into mechanical movement. Since, in accordance with accepted views, in this experiment the incandescence must be attributed to the impacts of the particles, molecules or atoms in tire heated space, these particles must therefore, in order to explain such action, be assumed to behave as independent carriers of electric charges immersed in an insulating medium; yet there is no attractive force between the glass tube and the filament because the space in the tube is, as a whole, conducting.

It is of some interest to observe in this connection that whereas the attraction between two electrified bodies may cease owing to the impairing of the insulating power of the medium in which they are immersed, the repulsion between the bodies may still be observed. This may be explained in a plausible way. When the bodies are placed at some distance in a poorly conducting medium, such as slightly warmed or rarefied air, and are suddenly electrified, opposite electric charges being imparted to them, these charges equalize more or less by leakage through the air. But if the bodies are similarly electrified, there is less opportunity afforded for such dissipation, hence the repulsion observed in such case is greater than the attraction. Repulsive actions in a gaseous medium are however, as Prof. Crookes has shown, enhanced by molecular bombardment.

ON CURRENT OR DYNAMIC ELECTRICITY PHENOMENA.

Fig. 11.

So far, I have considered principally effects produced by a varying electrostatic force in an insulating medium, such as air. When such a force is acting upon a conducting body of measurable dimensions, it causes within the same, or on its surface, displacements of the electricity, and gives rise to electric currents, and these produce another kind of phenomena, some of which I shall presently endeavor to illustrate. In presenting this second class of electrical effects, I will avail myself principally of such as are producible without any return circuit, hoping to interest you the more by presenting these phenomena in a more or less novel aspect.

It has been a long time customary, owing to .the limited experience with vibratory currents, to consider an electric current as something circulating in a closed conducting path. It was astonishing at first to realize that a current may flow through tile conducting path even if the latter be interrupted; and it was still more surprising to learn, that sometimes it may be even easier to make a current flow under such conditions than through a closed path. But that old idea is gradually disappearing, even among practical men, and will soon be entirely forgotten:

Fig. 12.

If I connect an insulated metal plate P, Fig. 11, to one of the terminals T of the induction coil by means of a wire, though this plate be verv well insulated, a current passes through the wire when the coil is set to work. First I wish to give you evidence that there *is* a current passing through the connecting wire. An obvious way of demonstrating this is to insert between the terminal of the coil and the insulated plate a very thin platinum or german silver wire w and bring the latter to incandescence or fusion by the current. This requires a rather large plate of else current impulses of very high potential and frequency. Another way is to take a coil C, Fig. 11, containing many turns of thin insulated wire and to insert the same in the path of the current to the plate. When I connect one of the ends of the coil to the wire leading to another insulated plate P_1, and its other end to the terminal T_1 of the induction coil, and set the latter to work, a current passes through the inserted coil C and the existence of the current may be made manifest in various ways. For instance, I insert an iron core i within the coil. The current being one of very high frequency, will, if it be of some strength, soon bring the iron core to a noticeably higher temperature, as the hysteresis and current losses are great with such high frequencies. One might take a core of sole size, laminated or not, it would matter little; but ordinary iron .wire 1/16-th or 1/8-th of an inch thick is suitable for the purpose. While the induction coil is working, a current traverses the inserted coil and only a few moments are sufficient to bring the iron wire i to an elevated temperature sufficient to soften the sealing wax .s and cause a paper washer p fastened by it to the iron wire to fall off. Put with the apparatus such as I have here, other, much more interesting, demonstrations of this kind can be made. I have a secondary s, Fig;. 12, of coarse wire, wound upon a coil similar to the first. In the preceding experiment the current through the coil C, Fig. 11, was very small, but there, being many turns a strong heating effect was, ncverthless, produced in the iron wire. Had I passed that current through a conductor in order to show the hcating of the latter, the current might have been too small to produce the effect desired. But with this coil provided with a secondary winding, I can now transform the feeble current of high tension which passes through the primary P into a strong secondary current of low tension. and this current wilt quite certainly do what I expect. In a small glass tube (t, Fig. 12), I have enclosed a coiled platinum wire, w, this merely in order to protect the wire. On each end of the glass tube is scaled a terminal of stout wire to which one of the ends of the platinum wire w, is connected. I join the terminals of the secondary coil to these terminals and insert the primary p, between the insulated plate P₁, and the terminal T₁, of the induction coil as before. The latter being set to work, instantly the platinum wire w is rendered incandescent and can be fused, even if it be very thick.

Fig. 13.

Instead of the platinum wire I now take an ordinary 50-volt 16 c p. lamp. When I set the induction coil in operation the lamp filament is brought to high incandescence. It is, however, not necessary to use the insulated plate, for the lamp (l Fig 13) is rendered incandescent even if the plate P1 be disconnected. The secondary may also he connected to the primary as indicated by the dotted line in Fig. 13, to do away more or less with the electrostatic induction or to modify the action otherwise.

Fig. 14a.
Fig. 14b.

I may here call attention to a number of interesting observations "with the lamp. First, I disconnect one of the terminals of the lamp from the secondary s.

When the induction coil plays, a glow is noted which fills the whole bulb. This glow is due to electrostatic induction. It increases when the bulb is grasped with the hated, and the capacity of the experimenter's body thus added to the secondary circuit. The secondary, in effect, is equivalent to a metallic coating, which would be placed near the primary. If the secondary, or its equivalent. the coating, were placed symmetrically. to the primary, the electrostatic induction would be nil under ordinary conditions, that is, when a primary return circuit is used, as both halves would neutralize each other. The secondary is in fact placed symmetrically to the primary, but the action of both halves of the latter, when only one of its ends is connected to the induction coil, is not exactly equal; hence electrostatic induction takes place, and hence the glow in the bulb. I can nearly equalize the action of both halves of the primary by connecting the other, free end of the same to the insulated plate, as in the preceding experiment. When the plate is connected, the glow disappears. With a smaller plate it would not entirely disappear and then it would contribute to the brightness of the filament when the secondary is closed, by warming the air in the bulb.

Fig. 15a.
Fig. 15b.

To demonstrate another interesting feature, I have adjusted the coils used in a certain way. I first connect both the terminals of the lamp to the secondary, one end of the primary being connected to the terminal T_1 of the induction coil and the other to tae insulated plate P, as before. When the current is turned on, the lamp glows brightly, as shown in Fig. 14b, in which C is a fine wire coil and s a coarse wire secondary wound upon it. If the insulated plate P_1 is disconnected, leaving one of the ends a of the primary insulated, the filament becomes dark or generally it diminishes in brightness (Fig. 14a). Connecting again the plate P_1 and raising the frequency of the current, I make the filament quite dark or barely red (Fig. 15b). Once more I will disconnect the plate. One will of course infer that when the plate is disconnected, the current through the primary will be weakened, that therefore the E. M. F. will fall in the secondary .s and that the brightness of the lamp will diminish. This might be the case and the result can be secured by an easy adjustment of the coils; also by varying the frequency and potential of the currents. But it is perhaps of greater interest to note, that the lamp increases in brightness when the plate is disconnected (Fig; 15a). In this case all the energy the primary receives is now sunk info it, like the charge of a battery in an ocean cable, but most of that energy is recovered through the secondary and used to light the lamp. The current traversing the primary is strongest at the end b which is connected to the terminal T, of the induction coil, and diminishes in strength towards the remote end a. But the dynamic inductive effect exerted upon the secondary s is now greater than before, when the suspended plate was connected to the primary. These results might have been produced by a number of causes. For instance, the plate P_1 being connected, the reaction from the coil C may be such as to diminish the potential at the terminal T_1 of the induction coil, and therefore weaken the current through the primary of the coil C. Or the disconnecting of the plate may diminish the capacity effect with relation to the primary of the latter coil to such an extent that the current through it is diminished, though the potential at the terminal T_1 of the induction coil may be the same or even higher. Or the result might have been produced by the change of please of the primary and secondary currents and consequent reaction. But the chief determining factor is the relation of the self-induction and capacity of coil C and plate P_1 and the frequency of the currents. The greater brightness of the filament in Fig. 15a. is, however, in part due to the heating of the rarefied gas in the lamp by electrostatic induction; which, as before remarked, is greater when the suspended plate is disconnected.

Still another feature of some interest I tray here bring to your attention. When the insulated date is disconnected and the secondary of the coil opened, by approaching. a small object to the secondary, but very small sparks can be drawn from it, showing that the electrostatic induction is stn all in this case. But upon the secondary being closed upon itself or through the lamp, the filament ,glowing brightly, strong sparks are obtained from the secondary. The electrostatic induction is now much greater, because the closed secondary determines a greater flow of current through the primary and principally through that half of it which is connected to the induction coil. If now the bulb be grasped with the hand, the capacity of the secondary with reference to the primary is augmented by the experimenter's body and the luminosity of the filament is increased, the Incandescence now being due partly to the flow of current through the filament and partly to the molecular bombardinent of the rarefied gas in the bulb.

Fig. 16.

The preceding experiments will have prepared one for the next following results of interest, obtained in the course of these investigations. Since I can pass a current through an insulated wire merely by connecting one of its ends to the source of electrical nergy, since I can induce by it another current, magnetize all iron core, and, in short, perform all operations as though a return circuit were used, clearly I can also drive a motor by the aid of only one wire. On a former occasion I have described a simple form of motor comprising a single exciting coil, an iron core and disc. Fig. 16 illustrates a modified way of operating such an alternate current motor by currents induced in a transformer connected to one lead, and several other arrangements of circuits for operating a certain class of alternating motors founded on the action of currents of differing phase. In view of the present state of the art it is thought sufficient to describe these arrangements in a few words only. The diagram, Fig. 16 II., shows a primary coil P, connected with one of its ends to the line L leading from a high tension transformer terminal T_1. In inductive relation to this primary P is a secondary s of coarse wire in the circuit of which is a coil C. The currents induced in the secondary energize the iron core i, which is preferably, but not necessarily, subdivided, and set the metal disc *d* in rotation. Such a motor M_2 as diagrammatically shown in Fig. 16 II., has been called a "magnetic lag motor", but this expression may be objected to by those who attribute the rotation of the disc to *eddy* currents circulating in minute paths when the core i is finally subdivided. In order to operate such a motor effectively on the plan indicated, the frequencies should not he too high, not more than four or five thousand, though the rotation is produced even with ten thousand per second, or more.

In Fig. 16 I., a motor M1 having two energizing circuits, A and B, is diagrammatically indicated. The circuit A is connected to the line L and in series with it is a primary P, which may have its free end connected to an insulated plate P_1, such connection being indicated by the dotted lines. The other motor circuit B is connected to the secondary s which is in inductive relation to the primary P. When the transformer terminal T_1 is alternately electrified, currents traverse the open line L and also circuit A and primary P. The currents through the latter induce secondary currents in the circuit S, which pass through the energizing coil B of the motor. The currents through the secondary S and those through the primary P differ in phase 90 degrees, or nearly so, and are capable of rotating an armature placed in inductive relation to the circuits A and B.

In Fig. 16 III., a similar motor M 3 with two energizing circuits A_1 and B_1 is illustrated. A primary P, connected with one of its ends to the line L has a secondary S, which is preferably wound for a tolerably high E. M. F., and to which the two energizing circuits of the motor are connected, one directly to the ends of the secondary and the other through a condenser C, by the action of which the currents traversing the circuit A_1 and B_1 are made to differ in phase.

In Fig. 16 IV., still another arrangement is shown. In this case two primaries P_1 and P_2 are connected to the line L, one through a condenser C of small capacity, and the other directly. The primaries are provided with secondaries S_1 and S_2 which are in series with the energizing circuits, A_2 and B_2 and a motor M_3 the condenser C again serving to produce the requisite difference in the phase of the currents traversing the motor circuits. As such phase motors with two or more circuits are now well known in the art, they have been here illustrated diagrammatically. No difficulty whatever is found in operating a motor in the manner indicated, or in similar ways; and although such experiments up to this day present only scientific interest, they may at a period not far distant, be carried out with practical objects in view.

It is thought useful to devote here a few remarks to the subject of operating devices of all kinds by means of only one leading wire. It is quite obvious, that when high-frequency currents are made use of, ground connections are — at least when the E. M. F. of the currents is great — better than a return wire. Such ground connections are objectionable with steady or low frequency currents on account of destructive chemical actions of the former and disturbing influences exerted by both on the neighboring circuits; but with high frequencies these actions practically do not exist. Still, even ground connections become, superfluous when the E. M. F. is very high, for soon a condition is reached, when the current may be passed more economically through open, than through closed, conductors. Remote as might seem an industrial application of such single wire transmission of energy to one not experienced in such lines of experiment, it will not seem so to anyone who for some time has carried on investigations of such nature. Indeed I cannot see why such a plan, should not be practicable. Nor should it be thought that for carrying out such a plan currents of very high frequency are expressly required, for just as soon as potentials of say 30,000 volts are used, the single wire transmission may be effected with low frequencies, and experiments have been made by me from which these inferences are trade.

Fig. 17.

When the frequencies are very high it has been found in laboratory practice quite easy to regulate the effects in the manner shown in diagram Fig. 17. Here two primaries P and P_1 are shown, each connected with one of its ends to the line L and with the other end to the condenser plates C and C, respectively. Near these are placed other condenser plates C_1 and C_1, the former being connected to the line L and the latter to an insulated larger plate P_2. On the primaries are wound secondaries S and S_1, of coarse wire, connected to the devices d and l respectively. By varying the distances of the condenser plates C and C_1, and C and C_1 the currents through the secondaries S and S_1 are varied in intensity. The curious feature is the great sensitiveness, the slightest change in the distance of the plates producing considerable variations in the intensity or strength of the currents. The sensitiveness may be rendered extreme by making the frequency such, that the primary itself, without any plate attached to its free end, satisfies, in conjunction with the closed secondary, the condition of resonance. In such condition an extremely small change in the capacity of the free terminal produces great variations. For instance, I have been able to adjust the conditions so that the mere approach of a person to the coil produces a considerable change in the brightness of the lamps attached to the secondary. Such observations and experiments possess, of course, at present, chiefly scientific interest, but they may soon become of practical importance.

Fig. 18.

Very high frequencies are of course not practicable with motors on account of the necessity of employing iron cores. But one may use sudden discharges of low frequency and thus obtain certain advantages of high-frequency currents without rendering the iron core entirely incapable of following the changes and without entailing a very great expenditure of energy in the core. I have found it quite practicable to operate with such low frequency disruptive discharges of condensers, alternating-current motors. A certain class of such motors which I advanced a few years ago, which contain closed secondary circuits, will rotate quite vigorously when the discharges are directed through the exciting coils: One reason that such a motor operates so well with these discharges is that the difference of phase between the primary and secondary currents is 90 degrees, which is generally not the case with harmonically rising and falling currents of low frequency. It might not be without interest to show an experiment with a simple motor of this kind, inasmuch as it is commonly thought that disruptive discharges are unsuitable for such purposes. The motor is illustrated in Fig. 18. It comprises a rather large iron core i with slots on the top into which are embedded thick copper washers C C. In proximity to the core is a freely-movable metal disc D. The core is provided with a primary exciting coil C_1 the ends a and b of which are connected to the terminals of the secondary S of an ordinary transformer, the primary P of the latter being connected to an alternating distribution circuit or generator G of low or moderate frequency. The terminals of the secondary S are attached to a condenser C which discharges through an air gap *d d* which may be placed in series or shunt to the coil C_1.When the conditions are properly chosen the disc D rotates Keith considerable effort and the iron core i does not get very perceptibly hot. With currents from a high-frequency alternator, on the contrary, the core gets rapidly hot and the disc rotates with a much smaller effort. To perform the experiment properly it should be first ascertained that the disc D is not set in rotation when the discharge is not occurring at *d d*. It is preferable to use a large iron core and a condenser of large capacity so as to bring the superimposed quicker oscillation to a very low pitch or to do away with it entirely. By observing certain elementary rules I have also found it practicable to operate ordinary series or shunt direct-current motors with such disruptive discharges, and this can be done with or without a return wire.

IMPEDANCE PHENOMENA.

Among the various current phenomena observed, perhaps the most interesting are those of impedance presented by conductors to currents varying at a rapid rate. In my first paper before the American Institute of Electrical Engineers, I have described a few striking observations of this kind. Thus I showed that when such currents or sudden discharges are passed through a thick metal bar there may be points on the bar only a few inches apart, which have a sufficient potential difference between them to maintain at bright incandescence an ordinary filament lamp. I have also described the curious behavior of rarefied gas surrounding a conductor, due to such sudden rushes of current. These phenomena have since been more carefully studied and one or two novel experiments of this kind are deemed of sufficient interest to be described here.

Figs. 19a, 19b and 19c.

Referring to Fig. 19a, B and B_1 are very stout copper bars connected at their lower ends to plates C and C_1, respectively, of a condenser, the opposite plates of the latter being connected to the terminals of the secondary S of a high-tension transformer, the primary P of which is supplied with alternating currents from an ordinary low-frequency dynamo G or distribution circuit. The condenser discharges through an adjustable gap d d as usual. By establishing a rapid vibration it was found quite easy to perform the following curious experiment. The bars B and B_1 were joined at the top by a low-voltage lamp l_3 a little lower was placed by means of clamps C C, a 50-volt lamp l_2; and still lower another 100-volt lamp l_1; and finally, at a certain distance below the latter lamp, an exhausted tube T. By carefully determining the positions of these devices it was found practicable to maintain them all at their proper illuminating power. Yet they were all connected in multiple arc to the two stout copper bars and required widely different pressures. This experiment requires of course some time for adjustment but is quite easily performed.

In Figs. 19b and 19c, two other experiments are illustrated which, unlike the previous experiment, do not require very careful adjustments. In Fig. 19b, two lamps, l1 and $l2$, the former a 100-volt and the latter a 50-volt are placed in certain positions as indicated, the 100-volt lamp being below the 50-volt lamp. When the arc is playing at d d and the sudden discharges are passed through the bars B B_1, the 50-volt lamp will, as a rule, burn brightly, or at least this result is easily secured, while the 100-volt lamp will burn very low or remain quite dark, Fig. 19b. Now the bars B B_1 may be joined at the top by a thick cross bar B_2 and it is quite easy to maintain the 100-volt lamp at full candle-power while the 50-volt lamp remains dark, Fig. 19c. These results, as I have pointed out previously, should not be considered to be due exactly to frequency but rather to the time rate of change which may be great, even with low frequencies. A great many other results of the same kind, equally interesting, especially to those who are *only* used to manipulate steady currents, may be obtained and they afford precious clues in investigating the nature of electric currents.

In the preceding experiments I have already had occasion to show some light phenomena and it would now be proper to study these in particular; but to make this investigation more complete I think it necessary to make first a few remarks on the subject of electrical resonance which has to be always observed in carrying out these experiments.

ON ELECTRICAL RESONANCE.

The effects of resonance are being more and more noted by engineers and are becoming of great importance in the practical operation of apparatus of all kinds with alternating currents. A few general remarks may therefore be made concerning these effects. It is clear, that if we succeed in employing the effects of resonance practically in the operation of electric devices the return wire will, as a matter of course, become unnecessary, for the electric vibration may be conveyed with one wire just as well as, and sometimes even better than, with two. The question first to answer is, then, whether pure resonance effects are producible. Theory and experiment both show that such is impossible in Nature, for as the oscillation becomes more and more vigorous, the losses in the vibrating bodies and environing media rapidly increase and necessarily check the vibration which otherwise would go on increasing forever. It is a fortunate circumstance that pure resonance is not producible, for if it were there is no telling what dangers might not lie in wait for the innocent experimenter. But to a certain degree resonance is producible, the magnitude of the effects being limited by the imperfect conductivity and imperfect elasticity of the media or, generally stated, by frictional losses. The smaller these losses, the more striking are the effects. The same is the case in mechanical vibration. A stout steel bar may be set in vibration by drops of water falling, upon it at proper intervals; and with glass, which is more perfectly elastic, the resonance effect is still more remarkable, for a goblet may be burst by singing into it a Tote of the proper pitch. The electrical resonance is the more perfectly attained, the smaller the resistance or the impedance of the conducting path and the more perfect the dielectric. In a Leyden jar discharging through a short stranded cable of thin wires these requirements are probably best fulfilled, and the resonance effects are , therefore very prominent. Such is not the case with dynamo machines, transformers and their circuits, or with commercial apparatus in general in which the presence of iron cores complicates the action or renders it impossible. In regard to Leyden jars with which resonance effects are frequently demonstrated, I would say that the effects observed are often *attributed* but are seldom *due* to true resonance, for an error is quite easily made in this respect. This may be undoubtedly demonstrated by the following experiment. Take, for instance, two large insulated metallic plates or spheres which I shall designate A and B; place them at a certain small distance apart and charge them from a frictional; or influence machine to a potential so high that just a slight increase of the difference of potential between them will cause the small air or insulating space to break down. This is easily reached by malting a few preliminary trials. If now another plate — fastened on an insulating handle and connected by a wire to one of the terminals of a high tension secondary of an induction coil, which is maintained in action by an alternator (preferably high frequency) — is approached to one of the charged bodies A or B, so a s to be nearer to either one of them, the discharge will invariably occur between them; at least it will, if the potential of the coil in connection with the plate is sufficiently high. But the explanation of this will soon be found in the fact that the approached plate acts inductively upon the bodies A and B and causes a spark to pass between them. When this spark occurs, the charges which were previously imparted to these bodies from the influence machine, must needs be lost, since the bodies are brought in electrical connection through the arc formed. Now this arc is formed whether there be resonance or not. But *even* if the spark would not be produced, still there is an alternating E. M. F. set up between the bodies when the plate is brought near one of them; therefore the approach of the plate, if it *does* not always actually, will, at any rate, *tend* to break down the air space by inductive action. Instead of the spheres or plates A and B we may take the coatings of a Leyden jar with the same result, and in place of the machine, — which is a high frequency alternator preferably, because it is more suitable for the experiment and also for the argument, — we may take another Leyden jar or battery of jars. When such jars are discharging through a circuit of low resistance the same is traversed by currents of very high frequency. The plate may now be connected to one of the coatings of the second jar, and when it is brought near to the first jar just previously charged to a high potential from an influence machine, the result is the same as before, and the first jar will discharge through a small air space upon the second being caused to discharge. But both jars and their circuits need not be tuned any closer than a basso profundo is to the note produced by a mosquito, as simall sparks will be produced through the air space, or at least the latter will be considerably more strained owing to the setting up of an alternating E. M. F. by induction, which takes place when one of the jars begins to discharge. Again another error of a similar nature is quite easily made. If the circuits of the two jars are run parallel and close together, and the experiment has been performed of discharging one by the other, and now a coil of wire be added to one of the circuits whereupon the experiment does not succeed, the conclusion that this is due to the fact that the circuits are now not tuned, would be far from being safe. For the two circuits act as condenser coatings and the addition of the coil to one of them is equivalent to bridging them, at the point where the coil is placed, by a small condenser, and the effect of the latter might be to prevent the spark from jumping through the discharge space by diminishing the alternating E. M. F. acting across the same. All these remarks, and many more which might be added but for fear of wandering too far from the subject, are made with the pardonable intention of cautioning the unsuspecting student, who might gain an entirely unwarranted opinion of his skill at seeing every experiment succeed; but they are in no way thrust upon the experienced as novel observations.

In order to make reliable observations of electric resonance effects it is very desirable, if not necessary, to employ an alternator giving currents which rise and fall harmonically, as in working with make and break currents the observations are not always trustworthy, since many phenomena, which depend on the rate of change, may be produced with widely different frequencies. Even when making such observations with an alternator one is apt to be mistaken. When a circuit is connected to an alternator there are an indefinite number of values for capacity and self-induction which, in conjunction, will satisfy the condition of resonance. So there are in mechanics an infinite number of tuning forks which will respond to a note of a certain pitch, or loaded springs which have a definite period of vibration. But the resonance will be most perfectly attained in that case in which the motion is effected with the greatest freedom. Now in mechanics, considering the vibration in the common medium — that is, air — it is of comparatively little importance whether one tuning fork be somewhat larger than another, because the losses in the air are not very considerable. One may, of course, enclose a tuning fork in an exhausted vessel and by thus reducing the air resistance to a minimum obtain better resonant action. Still the difference would not be very great. But it would make a great difference if the tuning fork were immersed in mercury. In the electrical vibration it is of enormous importance to arrange the conditions so that the vibration is effected with the greatest freedom. The magnitude of the resonance effect depends, under otherwise equal conditions, on the quantity of electricity set in motion or on the strength of the current driven through the circuit. But the circuit opposes the passage of .the currents by reason of its impedance and therefore, to secure the best action it is necessary to reduce the impedance to a minimum. It is impossible to overcome it entirely, but merely in part, for the ohmic resistance cannot be overcome. But when the frequency of the impulses is very great, the flow of the current is practically determined by self-induction. Now self-induction can be overcome by combining it with capacity. If the relation between these is such, that at the frequency used they annul each other, that is, have such values as to satisfy the condition of resonance, and the greatest quantity of electricity is made to flow through the external circuit, then the best result is obtained. It is simpler and safer to join the condenser in series with the self-induction. It is clear that in such combinations there will be, for a given frequency, and considering only the fundamental vibration, values which will give the best result, with the condenser in shunt to the self-induction coil; of course more such values than with the condenser in series. But practical conditions determine the selection. In the latter case in performing the experiments one may take a small self-induction and a large capacity or a small capacity and a large self-induction, but the latter is preferable, because it is inconvenient to adjust a large capacity by small steps. By taking a coil with a very large self-induction the critical capacity is reduced to a very small value, and the capacity of the coil itself may be sufficient. It is easy, especially by observing certain artifices, to wind a coil through which the impedance will be reduced to the value of the ohmic resistance only; and for any coil there is, of course, a frequency at which the maximum current will be made to pass through the coil. The observation of the relation between self-induction, capacity and frequency is becoming important in the operation of alternate current apparatus, such as transformers or motors, because by a judicious determination of the elements the employment of an expensive condenser becomes unnecessary. Thus it is possible to pass through the coils of an alternating current motor under the normal working conditions the required current with a low E. M. F. and do away entirely with the false current, and the larger the motor, the easier such a plan becomes practicable; but it is necessary for this to employ currents of very high potential or high frequency.

Fig. 20.

In Fig. 20 I. is shown a plan which has been followed in the study of ,the resonance effects by means of a high frequency alternator. C_1 is a coil of many turns, which is divided into small separate sections for the purpose of adjustment. The final adjustment was made sometimes with a few thin iron wires (though this is not always advisable) or with a closed secondary. The coil C_1 is connected with one of its ends to the line L from the alternator G and with the other end to one of the plates C of a condenser C C_1, the plate (C_1) of the latter being connected to a much larger plate P_1. In this manner both capacity and self-induction were adjusted to suit the dynamo frequency.

As regards the rise of potential through resonant action, of course, theoretically, it may amount to anything since it depends on self-induction and resistance and since these may have any value. But in practice one is limited in the selection of these values and besides these, there are other limiting causes. One may start with, say, 1,000 volts and raise the E. M. F. to 50 times that value, but one cannot start with 100,000 and raise it to ten times that value because of the losses in the media which are great, especially if the frequency is high. It should be possible to start with, for instance, two volts from a high or low frequency circuit of a dynamo and raise the E. M. F. to many hundred times that value. Thus coils of the proper dimensions might be connected each with only one of its ends to the mains from a machine of low E. M. F., and though the circuit of the machine would not be closed in the ordinary acceptance of the term, yet the machine might be burned out if a proper resonance effect would be obtained. I have not been able to produce, nor have

I observed with currents from a dynamo machine, such great rises of potential. It is possible, if not probable, that with currents obtained from apparatus containing iron the disturbing influence of the latter is the cause that these theoretical possibilities cannot be realized. But if such is the case I attribute it solely to the hysteresis and Foucault current losses in the core. Generally it was necessary to transform upward, when the E. M. F. was very low, and usually an ordinary form of induction coil was employed, but sometimes the arrangement illustrated in Fig. 20 II., has been found to be cenvenient. In this case a coil C is made in a great many sections, a few of these being used as a primary. In this manner both primary and secondary are adjustable. One end of the coil is connected to the line L_1 from the alternator, and the other line L is connected to the intermediate point of the coil. Such a coil with adjustable primary and secondary will be found also convenient in experiments with the disruptive discharge. When true resonance is obtained the top of the wave must of course be on the free end of the coil as, for instance, at the terminal of the phosphorescence bulb B. This is easily recognized by observing the potential of a point on the wire *w* near to the coil.

In connection with resonance effects and the problem of transmission of energy over a single conductor which was previously considered, I would *say a* few words on a subject which constantly fills my thoughts and which concerns the welfare of all. I mean the transmission of intelligible signals or perhaps even power to any distance without the use of wires. I am becoming daily more convinced of the practicability of the scheme; and though I know full well that the great majority of scientific men will not believe that such results can be practically and immediately realized, yet I think that all consider the developments in recent years by a number of workers to have been such as to encourage thought and experiment in this direction. My conviction has grown so strong, that I no loner look upon this plan of energy or intelligence transmission as a mere theoretical possibility, but as a serious problem in electrical engineering, which must be carried out some day. The idea of transmitting intelligence without wires is the natural outcome of the most recent results of electrical investigations. Some enthusiasts have expressed their belief that telephony to any distance by induction through the air is possible. I cannot stretch my imagination se far, but I do firmly believe that it is practicable to disturb by means of powerful machines the electrostatic condition of the carth and thus transmit intelligible signals and perhaps power. In fact, what is there against the carrying out of such a scheme? We now know that electric vibration may be transmitted through a single conductor. Why then not try to avail ourselves of the earth for this purpose? We need not be frightened by the idea of distance. To the weary wanderer counting the mile-posts the earth may appear very large but to that happiest of all men, the astronomer, who ,gazes at the heavens and by their standard judges the magnitude of our globe, it appears very small. And so I thinly it must seem to the electrician, for when he considers the speed with which an electric disturbance is propagated through the earth all his ideas of distance must completely vanish.

A point of great importance would be first to know what is the capacity of the earth? and what charge does it contain if electrified? Though we have no positive evidence of a charged body existing in space without other oppositely electrified bodies being near, there is a fair probability 'that the earth is such a body, for by whatever process it was separated from other bodies — and this is the accepted view of its origin — it must have retained a charge, as occurs in all processes of mechanical separation. If it be a charged body insulated in space its capacity should be extremely small, less than one-thousandth of a farad. But the upper strata of the air are conducting, and so, perhaps, is the medium in free space beyond the atmosphere, and these may contain an opposite charge. Then the capacity might be incomparably greater. In any case it is of the greatest importance to get an idea of what quantity of electricity the earth contains. It is difficult to say whether we shall ever acquire this necessary knowledge, but there is hope that we may, and that is, by means of electrical resonance. If ever we can ascertain at what period the earth's charge, when disturbed, oscillates with respect to an oppositely electrified system or known circuit, we shall know a fact possibly of the greatest importance to the welfare of the human race. I propose to seek for the period by means of an electrical oscillator, or a source of alternating electric currents. One of the terminals of the source would be connected to earth as, for instance, to the city water mains, the other to an insulated body of large surface. It is possible that the outer conducting air strata, or free space, contain an opposite charge and that, together with the earth, they form a condenser of very large capacity. In such case the period of vibration may be very low and an alternating dynamo machine might serve for the purpose of the experiment. I would then transform the current to a potential as high as it would be found possible and connect the ends of the high tension secondary to the ground and to the insulated body. By varying the frequency *of* the currents and carefully observing the potential of the insulated body and watching for the disturbance at various neighboring points of the earth's surface resonance might be detected. Should, as the majority of scientific men in all probability believe, the period be extremely small, then a dynamo machine would not do and a proper electrical oscillator would hate to be produced and perhaps it inight not be possible to obtain such rapid vibrations. But whether this be possible or not, and whether the earth contains a charge or not, and whatever may be its period of vibration, it certainly is possible — for of this we have daily evidence — to produce some electrical disturbance sufficiently powerful to be perceptible by suitable instruments at any point of the earth's surface.

Fig. 21.

Assume that a source of alternating currents be connected, as in Fig. 21, with one of its terminals to earth (conveniently to the water mains) and with the other to a body of large surface P. When the electric oscillation is set up there will be a movement of electricity in and out of P, and alternating currents will pass through the earth, converging to, or diverging from, the point C where the ground connection is trade. In this manner neighboring points on the earth's surface within a certain radius will be disturbed. But the disturbance will diminish with the distance, and the distance at which the effect will still be perceptible will depend on the quantity of electricity set in motion. Since the body P is insulated, in order to displace a considerable quantity, the potential of the source must be excessive, since there would be limitations as to the surface of P. The conditions might be adjusted so that the generator or source S will set up the swine electrical movement as though its circuit were closed. Thus it is certainly practicable to impress an electric vibration at least of a certain low period upon the earth by means of proper machinery. At what distance such a vibration might be made perceptible can only be conjectured. I have on another occasion considered the question how the earth might behave to electric disturbances. There is no doubt that, since in such an experiment the electrical density at the surface could be but extremely small considering the size of the earth, the air would not act as a very disturbing factor, and there would be not much energy lost through the action of the air, which would be the case if the density were great. Theoretically, then; it could not require a great amount of energy to produce a disturbance perceptible at great distance, or even all over the surface of the globe. Now, it is quite certain that at any point within a certain radius of the source S a properly adjusted self-induction and capacity device can be set in action by resonance. But not only can this be clone, but another source S_1 Fig. 21, similar to S, or any number of such sources, can be set to work in synchronism with the latter, and the vibration thus intensified and spread over a large area, or a flow of electricity produced to or from the source S_1 if the same be of opposite phase to the source S. I think that beyond doubt it is possible to operate electrical devices in a city through the ground or pipe system by resonance from an electrical oscillator located at a central point. But the practical solution of this problem would be of incomparably smaller benefit to man than the realization of the scheme of transmitting intelligence, or perhaps power, to any distance through the earth or environing medium. If this is at all possible, distance does not mean anything. Proper apparatus must first be produced by means of which the problem can be attacked and I have devoted much thought to this subject. I am firmly convinced that i: can be done and hope that we shall live to see it done.

ON THE LIGHT PHENOMENA PRODUCED BY HIGH-FREQUENCY CURRENTS OF HIGH POTENTIAL AND GENERAL REMARKS RELATING TO THE SUBJECT.

Returning now to the light effects which it has been the chief object to investigate, it is thought proper to divide these effects into four classes: 1. Incandescence of a solid. 2. Phosphorescence. 3. Incandescence or phosphorescence of a rarefied gas; and 4. Luminosity produced in a gas at ordinary pressure. The first question is: How are these luminous effects produced? In order to answer this question as satisfactorily as I am able to do in the light of accepted views and with the experience acquired, and to add some interest to this demonstration, I shall dwell here upon a feature which I consider of great importance, inasmuch as it promises, besides, to throw a better light upon the nature of most of the phenomena produced by high-frequency electric currents. I have on other occasions pointed out the great importance of the presence of the rarefied gas, or atomic medium in general, around the conductor through which alternate currents of high frequency are passed, as regards the heating of the conductor by the currents. My experiments, described some time ago, have shown that, the higher the frequency and potential difference of the currents, the more important becomes the rarefied gas in which the conductor is immersed, as a factor of the heating. The potential difference, however, is, as I then pointed out, a more important element than the frequency. When both of these are sufficiently high, the heating may be almost entirely due to the presence of the rarefied gas. The experiments to follow will show the importance of the rarefied gas, or, generally, of gas at ordinary or other pressure as regards the incandescence or other luminous effects produced by currents of this kind.

I take two ordinary 50-volt 16 C. P. lamps which are in every respect alike, with the exception, that one has been opened at the top and the air has filled the bulb, while the other is at the ordinary degree of exhaustion of commercial lamps. When I attach the lamp which is exhausted to the terminal of the secondary of the coil, which I have already used, as in experiments illustrated in Fig. 15a for instance, and turn on the current, the filament, as you have before seen, comes to high incandescence. When I attach the second lamp, which is filled with air, instead of the former, the filament still glows, but much less brightly. This experiment illustrates only in part the truth of the statements before made. The importance of the filament's being immersed in rarefied gas is plainly noticeable but not to such a degree as might be desirable. The reason is that the secondary of this coil is wound for low tension, having only 150 turns, and the potential difference at the terminals of the lamp is therefore small. Were I to take another coil with many more turns in the secondary, the effect would be increased, since it depends partially on the potential difference, as before remarked. But since the effect likewise depends on the frequency, it may be properly stated that it depends on the time rate of the variation of the potential difference. The greater this variation, the more important becomes the gas as an element of heating. I can produce a much greater rate of variation in another way, which, besides, has the advantage of doing away with the objections, which might be made in the experiment just shown, even if both the lamps were connected in series or multiple arc to the coil, namely, that in consequence of the reactions existing between the primary and secondary coil the conclusions are rendered uncertain. This result I secure by charging, from an ordinary transformer which is fed from the alternating current supply station, a battery of condensers, and discharging the latter directly through a circuit of small self-induction, as before illustrated in Figs. 19a, 19b and 19c.

Fig. 22a.   Fig. 22b.   Fig. 22c.

In Figs. 22a, 22b and 22c, the heavy copper bars BB₁ are connected to the opposite coatings of a battery of condensers, or generally in such way, that the high frequency or sudden discharges are made to traverse them. I connect first an ordinary 50-volt incandescent lamp to the bars by means of the clamps C C. The discharges being; passed through the lamp, the filament is rendered incandescent, though the current through it is very small, and would not be nearly sufficient to produce a visible effect under the conditions of ordinary use of the lamp. Instead of this I now attach to the bars another lamp exactly like the first, but with the seal broken off, the bulb being therefore filled with air at ordinary pressure. When the discharges are directed through the filament, as before, it does not become incandescent. But the result might still be attributed to one of the many possible reactions. I therefore connect both the lamps in multiple arc as illustrated in Fig. 22a. Passing tile discharges through both the lamps, again the filament in the exhausted

313

lamp *l* glows very brightly while that in the non-exhausted lamp l_1 remains dark, as previously. But it should not be thought that the latter lamp is' taking only a small fraction of the energy supplied to both the lamps; on the contrary, it may consume a considerable portion of the energy and it may become even hotter than the one which burns brightly. In this experiment the potential difference at the terminals of the lamps varies in sign theoretically three to four million times a second. The ends of the filaments are correspondingly electrified, and the gas in the bulbs is violently agitated and a large portion of the supplied energy is thus converted into heat. In the non-exhausted bulb, there being a few million times more gas molecules than in the exhausted one, the bombardment, which is most violent at the ends of the filament, in the neck of the bulb, consumes a large portion of the energy without producing any visible effect. The reason is that, there being many molecules, the bombardment is quantitatively considerable, but the individual impacts are not very violent, as the speeds of the molecules are comparatively small owing to the small free path. In the exhausted bulb, on the contary, the speeds are *very* great, and the individual impacts are violent and therefore better adapted to produce a visible effect. .Besides, the convection of heat is greater in the former bulb. In both the bulbs the current traversing the filaments is very small, incomparably smaller than that which they require on an ordinary low-frequency circuit. The potential difference, however, at the ends of the filaments is very great and might be possibly 20,000 volts or more, if the filaments were straight and their ends far apart. In the ordinary lamp a spark generally occurs between the ends of the filament or between the platinum wires outside, before such a difference of potential can be reached.

It might be objected that in the experiment before shown the lamps, being in multiple arc, the exhausted lamp might take a much larger current and that the effect observed might not be exactly attributable to the action of the gas in the bulbs. Such objections will lose much weight if I connect the lamps in series, with the wine result. When this is done and the discharges are directed through the filaments, it is again noted that the filament in the non-exhausted bulb l, remains dark, while that in the exhausted one (l) glows even snore intensely than under its normal conditions of working, Fig. 22b. According to general ideas the current through the filaments should now be the same, were it not modified by the presence of the gas around the filaments.

At this juncture I may point out another interesting feature, which illustrates the effect of the rate of change of potential of the currents. I will leave the two lamps connected in series to the bars BB_1, as in the previous experiment, Fig. 22b, but will presently reduce considerably the frequency of the currents, which was excessive in the experiment just before shown. This I may do by inserting a self-induction coil in the path of the discharges, or by augmenting the capacity of the condensers. When I now pass these low-frequency discharges through the lamps, the exhausted lamp l again is as bright as before, but it is noted also that the non-exhausted lamp l_1 glows, though not quite as intensely as the other. Reducing the current through the lamps, I may bring the filament in the latter lamp to redness, and, though the filament in the exhausted lamp I is bright, Fig. 22c, the degree of its incandescence is much smaller than in Fig. 22b; when the currents were of a much higher frequency.

In these experiments the gas acts in two opposite ways in determining the degree of the incandescence of the filaments, that is, by convection and bombardment. The higher the frequency and potential of the currents, the more important becomes the bomdardment. The convection on the contrary should be the smaller, the higher the frequency. When the currents are steady there is practically no bombardment, and convection may therefore with such currents also considerably modify the degree of incandescence and produce results similar to those just before shown. Thus, it two lamps exactly alike, one exhausted and one not exhausted, are connected in multiple arc or series to a direct-current machine, the filament in the non-exhausted lamp will require a considerably greater current to be rendered incandescent. This result is entirely due to convection, and the effect is the more prominent the thinner the filament. Professor Ayrton and Mr. Kilgour some time ago published quantitative results concerning the thermal emissivity by radiation and convection in which the effect with thin wires was clearly shown. This effect may be strikingly illustrated by preparing a number of small, short, glass tubes, each containing through its axis the thinnest obtainable platinum wire. If these tubes be highly exhausted, a number of them may be connected in multiple arc to a direct-current machine and all of the wires may be kept at incandescence with a smaller current than that required to render incandescent a single one of the wires if the tube be not exhausted. Could the tubes be so highly exhausted that convection would be nil, then the relative amounts of heat given off by convection and radiation could be determined without the difficulties attending thermal quantitative measurements. If a source of electric impulses of high frequency and very high potential is employed, a still greater number of the tubes may be taken and the wires rendered incandescent by a current not capable of warming perceptibly a wire of the same size immersed in air at ordinary pressure, and conveying the energy to all of them.

Fig. 23.

318

I may here describe a result which is still more interesting, and to which I have been led by the observation of these phenomena. I noted that small differences in the density of the air produced a considerable difference in the degree of incandescence of the wires, and I thought that, since in a tube, through which a luminous discharge is passed, tile gas is generally not of uniform density, a very thin wire contained in the tube might be rendered incandescent at certain places of smaller density of the gas, while it would remain dark at the places of greater density, where the convection would be greater and the bombardment less intense. Accordingly a tube t was prepared, as illustrated in Fig. 23, which contained through the middle a very fine platinum wire w. The tube was exhausted to a moderate degree and it was found that when it was attached to the terminal of a high-frequency coil the platinum wire w, would indeed, become incandescent in patches, as illustrated in Fig. 23. Later a number of these tubes with one or more wires were prepared, each showing this result. The effect was best noted when the striated discharge occurred in the tube, but was also produced when the striae were not visible, showing that, even then, the gas in the tube was, not or uniform density. The position of the striae was generally such, that the rarefactions corresponded to the places of incandescence or greater brightness on the wire w. But in a few instances it was noted, that the bright spots on the wire were covered by the dense parts of the striated discharge as indicated *by l* in Fig. 23, though the effect was barely perceptible. This was explained in a plausible way by assuming that the convection was not widely different in the

319

dense and rarefied places, and that the bombardment was greater on the dense places of the striated discharge. It is, in fact, often observed in bulbs, that under certain conditions a thin wire is brought to higher incandescence when the air is not too highly rarefied. This is the case when the potential of the coil is not high enough for the vacuum, but the result may be attributed to many different causes. In all cases this curious phenomenon of incandescence disappears when the tube, or rather the wire, acquires throughout a uniform temperature.

Disregarding now the modifying effect of convection there are then two distinct causes which determine the incandescence of a wire or filament with varying currents, that is, conduction current and bombardment. With steady currents we have to deal only with the former of these two causes, and the heating effect is a minimum, since the resistance is least to steady flow. When the current is a varying one the resistance is greater, and hence the heating effect is increased. Thus if the rate of change of the current is very great, the resistance may increase to such an extent that the filament is brought to incandescence with inappreciable currents, and we are able to take a short and thick block of carbon or other material and brim, it to bright incandescence with a current incomparably smaller than that required to bring to the same deuce of incandescence an ordinary thin lamp filament with a steady or low frequency current. This result is important, and illustrates how rapidly our views on these subjects are changing, and how quickly our field of knowledge is extending. In the art of incandescent lighting, to view this result in one aspect only, it has been commonly considered as an essential requirement for practical success, that the lamp filament should be thin and of high resistance. But now we know that the resistance of the filament to the steady flow does not mean anything; the filament might as well be short and thick; for if it be immersed in rarefied gas it will become incandescent by the passage of a small current. It all depends on the frequency and potential of the currents. We may conclude from this, that it would be of advantage, so far as the lamp is considered, to employ high frequencies for lighting, as they allow the use of short and thick filaments and smaller currents.

Fig. 24.

If a wire or filament be immersed in a homogeneous medium, all the heating is due to true conduction current, but if it be enclosed in an exhausted vessel the conditions are entirely different. Here the gas begins to act and the heating effect of the conduction current, as is shown in many experiments, may be *very* small compared with that of the bombardment. This is especially the case if the circuit is not closed and the potentials are of course *very* high. Suppose that a fine filament enclosed in an exhausted vessel be connected with one of its ends to the terminal of a high tension coil and with its other end to a large insulated plate. Though the circuit is not closed, the filament, as I have before shown, is brought to incandescence. If the frequency and potential be comparatively low, the filament is heated by the current passing *through it.* **If** the frequency and potential, and principally the latter, be increased, the insulated plate need be but very small, or may be done away with entirely; still: the filament will become incandescent, practically all the heating being then due to the bombardment. A practical way of combining both the effects of conduction currents and bombardment is Illustrated in Fig. 24, in which an ordinary lamp is shown provided with a very thin filament which has one of the ends of the latter connected to a shade serving the purpose of the insulated plate, and the other end to the terminal of a high tension source. It should not be thought that only rarefied gas is an important factor in the heating of a conductor by varying currents, but gas at ordinary pressure may become important, if the potential difference and frequency of the currents is excessive. On this subject I have already stated, that when a conductor is fused by a stroke of lightning, the current through it may be exceedingly small, not even sufficient to heat the conductor perceptibly, were. the latter immersed in a homogeneous medium.

From the preceding it is clear that when a conductor of high resistance is connected to the terminals of a source of high frequency currents of high potential, there may occur considerable dissipation of energy, principally at the ends of the conductor, in consequence of the action of the gas surrounding the conductor. Owing to this, the current through a section of the conductor at a point midway between its ends may be much smaller than through a section near the ends. Furthermore, the current passes principally through the outer portions of the conductor, but this effect is to be distinguished from the skin effect as ordinarily interpreted, for the latter would, or should, occur also in a continuous incompressible medium. If a great many incandescent lamps are connected in series to a source of such currents, the lamps at the ends may burn brightly, whereas those in the middle may remain entirely dark. This is due principally to bombardment, as before stated. But even if the currents be steady, provided the difference of potential is very great, the lamps at the end will burn more brightly than those in the middle. In such case there is no rhythmical bombardment, and the result is produced entirely by leakage. This leakage or dissipation into space when the tension is high, is considerable when incandescent lamps are used, and still more considerable with arcs, for the latter act like flames. Generally, of course, the dissipation is much smaller with steady, than with varying, currents.

I have contrived an experiment which illustrates in an interesting manner the effect of lateral diffusion. If a very long tube is attached to the terminal of a high frequency coil, the luminosity is greatest near the terminal and falls off gradually towards the remote end. This is more marked if the tube is narrow.

Fig. 25.

A small tube about one-half inch in diameter and twelve inches long (Fig. 25), has one of its ends drawn out into a fine fibre f nearly three feet long. The tube is placed in a brass socket T which can be screwed on the terminal T_1 of the induction coil. The discharge passing through the tube first illuminates the bottom of the same, which is of comparatively large section; but through the long glass fibre the discharge cannot pass. But gradually the rarefied gas inside becomes warmed and more conducting and the discharge spreads into the glass fibre. This spreading is so slow, that it may take half a minute or more until the discharge has worked through up to the top of the glass fibre, then presenting the appearance of a strongly luminous thin thread. By adjusting the potential at the terminal the light may be made to travel upwards at any speed. Once, however, the glass fibre is heated, the discharge breaks through its entire length instantly. The interesting point to be noted is that, the higher the frequency of the currents, or in other words, the greater relatively the lateral dissipation, at a slower rate may the light be made to propagate through the fibre. This experiment is best performed with a highly exhausted and freshly made tube. When the tube has been used for some time the experiment often fails. It is possible that the gradual and slow impairment of the vacuum is the cause. This slow propagation of the discharge through a very narrow glass tube corresponds exactly to the propagation of heat through a bar warmed at one end. The quicker the heat is carried away laterally the longer time it will take for the heat to warm the remote end. When the current of a low frequency coil is passed through the fibre from end to end, then the lateral dissipation is small and the discharge instantly breaks through almost without exception.

Fig. 26.

After these experiments and observations which have shown the importance of the discontinuity or atomic structure of the medium and which will serve to explain, in a measure at least, the nature of the four kinds of light effects producible with these currents,

I may now give you an illustration of these effects. For the sake of interest I may do this in a manner which to many of you might be novel. You have seen: before that we may now convey the electric vibration to a body by means of a single wire or conductor of any hind. Since the human frame is conducting I may convey the vibration through my body.

Fig. 27.

First, as in some previous experiments, I connect my
body with one of the terminals of a high-tension

transformer and take in my hand an exhausted bulb which contains a small carbon button mounted upon a platinum wire leading to the outside of the bulb, and the button is rendered incandescent as soon as the transformer is set to work (Fig. 26). I may place a conducting shade on the bulb which serves to intensify the action, but it is not necessary. Nor is it required that the button should be in conducting connection with the hand through a wire leading through the glass, for sufficient energy may be transmitted through the glass itself by inductive action to render the button incandescent.

Fig. 28.

Next I take a highly exhausted bulb containing a strongly phosphorescent body, above which is mounted a small plate of aluminum on a platinum wire leading to the outside, and the currents flowing through my body excite intense phosphorescence in the bulb (Fig. 27). Next again I take in my hand a simple exhausted tube, and in the same manner the gas inside the tube is rendered highly incandescent or phosphorescent (Fig. 28). Finally, I may take in my hand a wire, bare or covered with thick insulation, it is quite immaterial; the electrical vibration is so intense as to cover the wire with a luminous film (Fig. 29).

Fig. 29.

A few words must now be devoted to each of these phenomena. In the first place, I will consider the incandescence of a button or of a solid in general, and dwell upon some facts which apply equally to all these phenomena. It was pointed out before that when a thin conductor, such as a lamp filament, for instance, is connected with one of its ends to the terminal of a transformer of high tension the filament is brought to incandescence partly by a conduction current and partly by bombardment. The shorter and thicker the filament the more important becomes the latter, and finally, reducing the filament to a mere button, all the heating must practically be attributed to the bombardment. So in the experiment before shown, the button is rendered incandescent by the rhythmical impact of freely movable small bodies in the bulb. These bodies may be the molecules of the residual gas, particles of dust or lumps torn from the electrode; whatever they are, it is certain that the heating of the button is essentially connected with the pressure of such freely movable particles, or of atomic matter in general in the bulb. The heating is the more intense the greater the number of impacts per second and the greater the energy of each impact. Yet the button would be heated also if it were connected to a source of a steady potential. In such a case electricity mould be carried away from the button by the freely movable carriers or particles flying about, and the quantity of electricity thus carried away might be sufficient to bring the button to incandescence by its passage through the latter. But the bombardment could not be of great importance in such case. For this reason it would require a comparatively very great supply of energy to the button to maintain it at incandescence with a steady potential. The higher the frequency of the electric impulses the more economically can the button be maintained at incandescence. One of the chief reasons why this is so, is, I believe, that with impulses of very high frequency there is less exchange of the freely movable carriers around the electrode and this means, that in the bulb the heated matter is better confined to the neighborhood of the button. If a double bulb, as illustrated in Fig. 30 be made, comprising a large globe B and a small *one b,* each containing as usual a filament *f* mounted on a platinum wire *w and* w_1 *it* is found, that if the filaments *f f* be exactly alike, it requires less energy to keep the filament in the globe b at a certain degree of incandescence, than that in the globe B. This is due to the confinement of the movable particles around the button. In this case it is also ascertained, that the filament in the small globe b is less deteriorated when maintained a certain length of time at incandescence. This is a necessary consequence of the fact that the gas in the small bulb becomes strongly heated and therefore a very good conductor, and less work is then performed on the button, since the bombardment becomes less intense as the conductivity of the gas increases. In this construction, of course, the small bulb becomes very hot and when it reaches an elevated temperature the convection and radiation on the outside increase. On another occasion I have shown bulbs in which this drawback was largely avoided. In these instances a very small bulb, containing a refractory button, was mounted in a large globe and the space between the walls of both was highly exhausted. The outer large globe remained comparatively cool in such constructions, When the large globe was on the pump

336

and the vacuum between the walls maintained permanent by the continuous action of the pump, the outer globe would remain quite cold, while the button in tile small bulb was kept at incandescence. But when the seal was made, and the button in the small bulb maintained incandescent some length of time, the large globe too would become warmed. Prom this I conjecture that if vacuous space (as Prof. Dewar finds) cannot convey heat, it is so merely in virtue of our rapid motion through space or, generally speaking, by the motion of the medium relatively to us, for a permanent condition could not be maintained without the medium being constantly renewed. A vacuum cannot, according to all evidence, be permanently maintained around a hot *body.*

Fig. 30.

In these constructions, before mentioned, the small bulb inside would, at least in the first stages, prevent all bombardment against the outer large globe. It occurred to me then to ascertain how a metal sieve would behave in this respect, and several bulbs, as illustrated in Fig. 31, were prepared for this purpose.

In a globe b, was mounted a thin filament *f* (or button) upon a platinum wire w passing through a glass stem and leading to the outside of the globe. The filament f was surrounded by a metal sieve s. It was found in experiments with such bulbs that a sieve with wide meshes apparently did not in the slightest affect the bombardment against the globe b. When the vacuum was high, the shadow of the sieve was clearly projected against the globe and the latter would get hot in a short while. In some bulbs the sieve ,v was connected to a platinum wire sealed in the glass. When this wire was connected to the other terminal of the induction coil (the E. M. F. being kept low in this case), or to an insulated plate, the bombardment against the outer globe b was diminished. By taking a sieve with fine meshes the bombardment against the globe b was always diminished, but even then if the exhaustion was carried very far, and when the potential of the transformer was very high, the globe b would be bombarded and heated quickly, though no shadow of the sieve was visible, owing to the smallness of the meshes. But a glass tube or other continuous *body* mounted so as to surround the filament, did entirely cut off the bombardment and for a while the outer globe b would remain perfectly cold. Of course when the glass tube was sufficiently heated the bombardment against the outer globe could be noted at once. The experiments with these bulbs seemed to show that the speeds of the projected molecules or particles must be considerable (though quite insignificant when compared with that of light), otherwise it would be difficult to understand how *they* could traverse a fine metal sieve without

339

being affected, unless it were found that such small particles or atoms cannot be acted upon directly at measurable distances. In regard to the speed of the projected atoms, Lord Kelvin has recently estimated it at about one kilometre a second or thereabouts in an ordinary Crookes bulb. As the potentials obtainable with a disruptive discharge coil are much higher than with ordinary coils, the speeds must, of course, be much greater when the bulbs are lighted from such a coil. Assuming the speed to be as high as five kilometres and uniform through the whole trajectory, as it should be in a very highly exhausted vessel, then if the alternate electrifications of the electrode would be of a frequency of five million, the greatest distance a particle could get away from the electrode would be one millimetre, and if it could be acted upon directly at that distance, the exchange of electrode matter or of the atoms would be very slow and there would be practically no bombardment against the bulb. This at least should be so, if the action of an electrode upon the atoms of the residual gas would be such as upon electrified bodies which we can perceive. A hot body enclosed in a n exhausted bulb produces always atomic bombardment, but a hot body has no definite rhythm, for its molecules perform vibrations of all kinds.

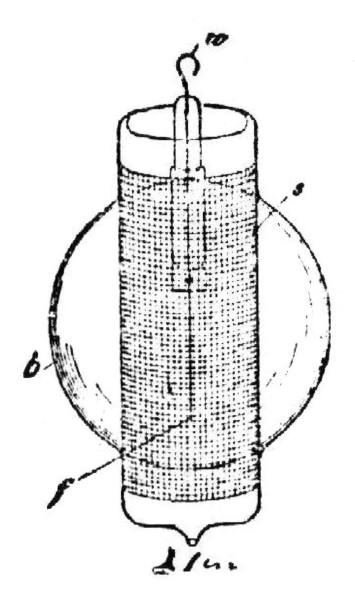

Fig. 31.

If a bulb containing a button or filament be exhausted
as high as is possible with tile greatest care and by
the use of the best artifices, it is often observed that

the discharge cannot, at first, break through, but after some time, probably in consequence of some changes within the bulb, the discharge finally passes through and the button is rendered incandescent. In fact, it appears that the higher the degree of exhaustion the easier is the incandescence produced. There seem to be no other causes to which the incandescence might be attributed in such case except to the bombardment or similar action of the residual gas, or of particles of matter in general. But if the bulb be exhausted with the greatest care can these play an important part? Assume the vacuum in the bulb to be tolerably perfect, the great interest then centres in the question: Is the medium which pervades all space continuous or atomic? If atomic, then the heating of a conducting button or filament in an exhausted vessel might be due largely to ether bombardment, and then the heating of a conductor in general through which currents of high frequency or high potential are passed must be modified by the behavior of such medium; then also the skin effect, the apparent increase of the ohmic resistance, etc., admit, partially at least, of a different explanation.

It is certainly more in accordance with many phenomena observed with highfrequency currents to hold that all space is pervaded with free atoms, rather than to assume that it is devoid of these, and dark and cold, for so it must be, if filled with a continuous medium, since in such there can be neither heat nor light. Is then energy transmitted by independent carriers or by the vibration of a continuous medium? This important question is by no means as yet positively answered. But most of the effects which are here considered, especially the light effects, incandescence, or phosphorescence, involve the presence of free atoms and would be impossible without these.

In regard to the incandescence of a refractory button (or filament) in an exhausted receiver, which has been one of the subjects of this investigation, the chief experiences, which may serve as a guide in constructing such bulbs, may be summed up as follows: 1. The button should be as small as possible, spherical, of a smooth or polished surface, and of refractory material which withstands evaporation best. 2. The support of the button should be very thin and screened by an aluminum and mica sheet, as I have described on another occasion. 3. The exhaustion of the bulb should be as high as possible. 4. The frequency of the currents should be as high as practicable. 5. The currents should be of a harmonic rise and fall, without sudden interruptions. 6. The heat should be confined to the button by inclosing the same in a small bulb or otherwise. 7. The space between the walls of the small bulb and the outer globe should be highly exhausted.

Most of the considerations which apply to the incandescence of a solid just considered may likewise be applied to phosphorescence. Indeed, in an exhausted vessel the phosphorescence is, as a rule, primarily excited by the powerful beating of the electrode stream of atoms against the phosphorescent body. Even in many cases, where there is no evidence of such a bombardment, I think that phosphorescence is excited by violent impacts of atoms, which are not necessarily thrown off from the electrode but are acted upon from the same inductively through the medium or through chains of other atoms. That mechanical shocks play an important part in exciting phosphorescence in a bulb may be seen from the following experiment. If a bulb, constructed as that illustrated in Fig. 10, be taken and exhausted with the greatest care so that the discharge cannot pass, the filament f acts by electrostatic induction upon the tube t and the latter is set in vibration. If the tube o be rather wide, about an inch or so, the filament may be so powerfully vibrated that whenever it hits the glass tube it excites phosphorescence. But the phosphorescence ceases when the filament comes to rest. The vibration can be arrested and again started by varying the frequency of the currents. Now the filament has its own period of vibration, and if the frequency of the currents is such that there is resonance, it is easily set vibrating, though the potential of the currents be small. I have often observed that the filament in the bulb is destroyed by such mechanical resonance. The filament vibrates as a rule so rapidly that it cannot be seen and the experimenter may at first be mystified. When such an experiment as the one described is carefully performed, the potential of the currents need be extremely small, and for this reason I infer that the phosphorescence is then due to the mechanical shock of the filament against the glass, just as it is produced by striking a loaf of sugar with a knife. The mechanical shock produced by the projected atoms is easily noted when a bulb containing a button is grasped in the hand and the current turned on suddenly. I believe that a bulb could be shattered by observing the conditions of resonance.

In the experiment before cited it is, of course, open to say, that the glass tube, upon coming in contact with the filament, retains a charge of a certain sign upon the point of contact. If now the filament main touches the glass at the same point while it is oppositely charged, the charges equalize under evolution of light. But nothing of importance would be gained by such an explanation. It is unquestionable that the initial charges given to the atoms or to the glass play some part in exciting phosphorescence. So for instance, if a phosphorescent bulb be first excited by a high frequency coil by connecting it to one of the terminals of the latter and the degree of luminosity be noted and then the bulb be highly charged from a Holtz machine by attaching it preferably to the positive terminal of the machine, it is found that when the bulb is again connected to the terminal of the high frequency coil, the phosphorescence is far more intense. On another occasion I have considered the possibility of some phosphorescent phenomena in bulbs being produced by the incandescence of an infinitesimal layer on the surface of the phosphorescent body.

Certainly the impact of the atoms is powerful enough to produce intense incandescence by the collisions, since they bring quickly to a high temperature a body of considerable bulk. If any such effect exists, then the best appliance for producing phosphorescence in a bulb, which we know so far, is a disruptive discharge coil giving an enormous potential with but few fundamental discharges, say 25-30 per second, just enough to produce a continuous impression upon the eye. It is a fact that such a coil excites phosphorescence under almost any condition and at all degrees of exhaustion, and I have observed effect; which appear to be due to phosphorescence even at ordinary pressures of the atmosphere, when the potentials are extremely high. But if phosphorescent light is produced by the equalization of charges of electrified atoms (whatever this may mean ultimately), then the higher the frequency of the impulses or alternate electrifications, the more economical will be the light production. It is a long known and noteworthy fact that all the phosphorescent bodies are poor conductors of electricity and heat, and that all bodies cease to emit phosphorescent light when they are brought to a certain temperature. Conductors on the contrary do not possess this quality. There are but few exceptions to the rule. Carbon is one of them. Becquerel noted that carbon phosphoresces at a certain elevated temperature preceding the dark red. This phenomenon may be easily observed in bulbs provided with a rather large carbon electrode (say, a sphere of six millimetres diameter). If the current is turned on after a few seconds, a snow white film covers the electrode, just before it bets dark red. Similar effects are noted with other conducting bodies, but many scientific men will probably not attribute them to true phosphorescence. Whether true incandescence has anything to do with .phosphorescence excited by atomic impact or mechanical shocks still remains to be decided, but it is a fact that all conditions, which tend to localize and increase the heating effect at the point of impact, are almost invariably the most favorable for the production of phosphorescence. So, if the electrode be very small, which is equivalent to saying in general, that the electric density is great; if the potential be high, and if the gas be highly rarefied, all of which things imply high speed of the projected atoms, or matter, and consequently violent impacts — the phosphorescence is very intense. If a bulb provided with a large and small electrode be attached to the terminal of an induction coil, the small electrode excites phosporescence while the large one may not do so, because of the smaller electric density and hence smaller speed of the atoms. A bulb provided with a large electrode may be grasped with the hand while the electrode is connected to the terminal of the coil and it may not phosphoresce; but if instead of grasping the bulb with the hand, the same be touched with a pointed wire, the phosphorescence at once spreads through the bulb, because of the great density at the point of contact. With low frequencies it seems that gases of great atomic weight excite more intense phosphorescence than those of smaller weight, as for instance, hydrogen. With high frequencies the observations are not sufficiently reliable to draw a conclusion. Oxygen, as is well-known, produces exceptionally strong effects, which may be in part due to chemical action. A bulb with hydrogen residue seems to be most easily excited. Electrodes which are most easily deteriorated produce more intense phosphorescence in bulbs, but the condition is not permanent because of the impairment of the vacuum and the deposition of the electrode matter upon the phosphorescent surfaces. Some liquids, as oils, for instance, produce magnificent effects of phosphorescence (or fluorescence?), but they last only a few seconds. So if a bulb has a trace of oil on the walls and the current is turned on, the phosphorescence only persists for a few moments until the oil is carried away. Of all bodies so far tried, sulphide of zinc seems to be the most susceptible to phosphorescence. Some samples, obtained through the kindness of Prof. Henry in Paris, were employed in many of these bulbs. One of the defects of this sulphide is, that it loses its quality of emitting light when brought to a temperature which is by no means high. It can therefore, be used only for feeble intensities. An observation which might deserve notice is, that when violently bombarded from an aluminum electrode it assumes a black color, but singularly enough, it returns to the original condition when it cools down.

The most important fact arrived at in pursuing investigations in this direction is, that in all cases it is necessary, in order to excite phosphorescence with a minimum amount of energy, to observe ceitain conditions. Namely, there is always, no matter what the frequency of the currents, degree of exhaustion and character of the bodies in the bulb, a certain potential (assuming the bulb excited from one terminal) or potential difference (assuming the bulb to be excited with both terminals) which produces the most economical result. If the potential be increased, considerable energy may be wasted without producing any more light, and if it be diminished, then again the light production is not as economical.

The exact condition under which the best result is obtained seems to depend on many things of a different nature, and it is to be yet investigated by other experimenters, but it will certainly have to be observed when such phosphorescent bulbs are operated, if the best results are to be obtained.

Fig. 32.

Coming now to the most interesting of these phenomena, the incandescence or phosphorescence of gases, at low pressures or at the ordinary pressure of the atmosphere, we must seek the explanation of these phenomena in the same primary causes, that is, in shocks or impacts of the atoms. Just as molecules or atoms beating upon a solid body excite phosphorescence in the same or render it incandescent, so when colliding among themselves they produce similar phenomena. But this is a very insufficient explanation and concerns only the crude mechanism. Light is produced by vibrations which go on at a rate almost inconceivable. If we compute, from the energy contained in the form of known radiations in a definite space the force which is necessary to set up such rapid vibrations, we find, that though the density of the ether be incomparably smaller than that of any body we know, even hydrogen, the force is something surpassing comprehension. What is this force, which in mechanical measure may amount to thousands of tons per square inch? It is electrostatic force in the light of modern views. It is impossible to conceive how a body of measurable dimensions could be charged to so high a potential that the force would be sufficient to produce these vibrations. Long before any such charge could be imparted to the body it would be shattered into atoms. The sun emits light and heat, and so does an ordinary flame or incandescent filament, but in neither of these can the force be accounted for if it be assumed that it is associated with the body as a whole. Only in one way may we account for it, namely, by identifying it with the atom. An atom is so small, that if it be charged by coming in contact with an electrified body and the charge be assumed to follow the same law as in the case of bodies of measurable dimensions, it must retain a quantity of electricity which is fully capable of accounting for these forces and tremendous rates of vibration. But the atom behaves singularly in this respect — it always takes the same "charge".

It is very likely that resonant vibration plays a most important part in all manifestations of energy in nature. Throughout space all matter is vibrating, and all rates of vibration are represented, from the lowest musical note to the highest pitch of the chemical rays, hence an atom, or complex of atoms, no matter what its period, must find a vibration with which it is in resonance. When we consider the enormous rapidity of the light vibrations, we realize the impossibility of producing such vibrations directly with any apparatus of measurable dimensions, and we are driven to the only possible means of attaining the object of setting up waves of light by electrical means and economically, that is, to affect the molecules or atoms of a gas, to cause them to collide and vibrate. We then must ask ourselves — How can free molecules or atoms be affected ?

Fig. 33.

It is a fact that they can be affected by electrostatic force, as is apparent in many of these experiments. By varying the electrostatic force we can agitate the atoms, and cause them to collide accompanied by evolution of heat and light. It is not demonstrated beyond doubt that vie can affect them otherwise. If a luminous discharge is produced in a closed exhausted tube, do the atoms arrange themselves in obedience to any other but to electrostatic force acting in straight lines from atom to atom? Only recently I investigated the mutual action between two circuits with extreme rates of vibration. When a battery of a few jars (c c c c, Fig. 32), is discharged through a primary P of low resistance (the connections being as illustrated in Figs. 19a, 19b and 19c), and the frequency of vibration is many millions there are great differences of potential between points on the primary not more than a few inches apart. These differences may be 10,000 volts per inch. if not more, taking the maximum value of the h. M. F. The secondary S is therefore acted upon by electrostatic induction, which is in such extreme cases of much greater importance than the electro-dynamic. To such sudden impulses the primary as well as the secondary are poor conductors, and therefore great differences of potential may be produced by electrostatic induction between adjacent points on the secondary. Then sparks may jump between the wires and streamers become visible in the dark if the light of the discharge through the spark gap d d be carefully excluded. If now we substitute a closed vacuum tube for the metallic secondary S, the differences of potential produced in the tube by electrostatic induction from the primary are fully sufficient to excite portions of it; but as the points of certain differences of potential on the primary are not fixed, but are generally constantly changing in position, a luminous band is produced in the tube, apparently not touching the glass, as it should, if the points of maximum and minimum differences of potential were fixed on the primary. I do not exclude the possibility of such a tube being excited only by electro-dynamic induction, for very able physicists hold this view; but in my opinions, there is as yet no positive proof given that atoms of a gas in a closed tube may arrange themselves in chains under the action of an: electromotive impulse produced by electro-dynamic induction in the tube. I have been unable so far to produce striae in a tube, however long, and at whatever degree of exhaustion, that is, striae at right angles to the supposed direction of the discharge or the axis of the tube; but I have distinctly observed in a large bulb, in which a wide luminous band was produced by passing a discharge of a battery through a wire surrounding the bulb, a circle of feeble luminosity between two luminous bands, one of which was more intense than the other. Furthermore, with my present experience I do not think that such a gas discharge in a closed tube can vibrate, that is, vibrate as a whole. I am convinced that no discharge through a gas can vibrate. The atoms of a gas behave very curiously in respect to sudden electric impulses. The gas does not seem to possess any appreciable inertia to such impulses, for it is a fact, that the higher the frequency of the impulses, with the greater freedom does the discharge pass through the gas. If the gas possesses no inertia then it cannot vibrate, for some inertia is necessary for the free vibration. I conclude from this that if a lightning discharge occurs between two clouds, there can be no oscillation, such as would be expected, considering the capacity of the clouds. But if the lightning discharge strike the earth, there is always vibration — in the earth, but not in the cloud. In a gas discharge each atom vibrates at its oven rate, but there is no vibration of the conducting gaseous mass as a whole. This is an important consideration in the great problem of producing light economically, for it teaches us that to reach this result we must use impulses of very high frequency and necessarily also of high potential. It is , a fact that oxygen produces a more intense light in a tube. Is it because oxygen atoms possess some inertia and the vibration does not die out instantly? But then nitrogen should be as good, and chlorine and vapors of many other bodies much better than *oxygen,* unless the magnetic properties of the latter enter prominently into play. Or, is the process in the tube of an electrolytic nature? Many observations certainly speak for it, the most important being that matter is always carried away from the electrodes and the vacuum in a bulb cannot be permanently maintained. If such process takes place in reality, then again must we take refuge in high frequencies, for, with such, electrolytic action should be reduced to a minimum, if not rendered entirely impossible. It is an undeniable fact that with very high frequencies, provided the impulses be of harmonic nature, like those obtained from an alternator, there is less deterioration and the vacua are more permanent. With disruptive discharge coils there are sudden rises of potential and the vacua are more quickly impaired, for the electrodes are deteriorated in a very short time. It was observed in some large tubes, which were provided with heavy carbon blocks B B_1, connected to platinum wires w w_1(as illustrated in Fig. 33), and which were employed in experiments with the disruptive discharge instead of the ordinary air gap, that the carbon particles under the action of the powerful magnetic field in which the tube was placed, were deposited in regular fine lines in the middle of the tube, as illustrated. These lines were attributed to the deflection or distortion of the discharge by the magnetic field, but why the deposit occurred principally where the field was most intense did not appear quite clear. A fact of interest, likewise noted, was that the presence of a strong magnetic field increases the deterioration of the electrodes, probably by reason of the rapid interruptions it produces, whereby there is actually a higher E. M. F. maintained between the electrodes. Much would remain to be said about the luminous effects produced in gases at low or ordinary pressures. With the present experiences before us we cannot say that the essential nature of these charming phenomena is sufficiently known. But investigations in this direction are being pushed with exceptional ardor. Every line of scientific pursuit has its fascinations, but electrical investigation appears to possess a peculiar attraction, for there is no experiment or observation of any kind in the domain of this wonderful science which would not forcibly appeal to us. Yet to me it seems, that of all the many marvelous thins we observe, ti vacuum tube, excited by an electric impulse from a distant source, bursting forth out of the darkness and illuminating the room with its beautiful light, is as lovely' a phenomenon as can greet our eyes. More interesting still it appears when, reducing the fundamental discharges across the gap to a very small number and waving the tube about we produce all hinds of designs in luminous lines. So by way of amusement I take a straight long tube, or a square one, or a square attached to a straight tube, and by whirling them about in the hand, I imitate the spokes of a wheel, a Gramme winding, a drum winding, an alternate current motor winding, etc (Fig. 34). Viewed from a distance the effect is weak and much of its beauty is lost, but being near or holding the tube in the hand, one cannot resist its charm.

Fig. 34.

In presenting these insignificant results I have not attempted to arrange and coordinate them, as would be proper in a strictly scientific investigation, in which every succeeding result should be a logical sequence of the preceding, so that it might be guessed in advance by the careful reader or attentive listener. I have preferred to concentrate my energies chiefly upon advancing novel facts or ideas which might serve as suggestions to others, and this may serve as an excuse for the lack of harmony. The explanations of the phenomena have been given in good faith and in the spirit of a student prepared to find that they admit of a better interpretation. There can be no great harp t in a student taking an erroneous view, but when great minds err, the world must dearly pay for their mistakes.

CPSIA information can be obtained
at www.ICGtesting.com
Printed in the USA
BVHW071718090321
602118BV00012B/1311